高等职业教育工匠工坊新型活页式系列教材

云计算
应用运维实战

郭立文　魏晓艳　王海龙◎主　编
夏卓越　宁春梅　应天龙◎副主编

中国铁道出版社有限公司

2023年·北京

内 容 简 介

本书是云计算技术入门书籍，适合初学者使用。全书共 6 个项目。项目 1 主要介绍自动化运维，内容包括 Ansible 安装以及 Ansible Playbook 使用；项目 2 主要介绍 Docker 搭建与运维，内容包括 Docker 的安装和基本操作；项目 3 主要介绍 ELK 日志分析系统，内容包括 ELK 的安装和使用、Logstash 部署以及收集 Nginx 日志；项目 4 主要介绍 MySQL 常用操作；项目 5 主要介绍网站优化与服务器优化；项目 6 主要介绍 Tomcat 搭建配置。

全书参照开发云计算功能所需要的技术，如 Ansible 与 Ansible Playbook 的使用、Docker 容器帮助快速部署、Elasticsearch 日志分析工具、网站优化与服务器优化，对云计算的设计和开发过程进行了系统而又详细的介绍，使读者能清楚地了解云计算开发过程中的各种知识。本书附有相关代码和界面截图，希望帮助读者更好地学习。

本书适合作为高等职业院校计算机类相关专业的教材，也可作为编程爱好者的参考用书。

图书在版编目（CIP）数据

云计算应用运维实战/郭立文，魏晓艳，王海龙主编.—北京：中国铁道出版社有限公司，2023.5
高等职业教育工匠工坊新型活页式系列教材
ISBN 978-7-113-29943-9

I.①云… II.①郭…②魏…③王… III.①云计算-高等职业教育-教材 IV.①TP393.027

中国国家版本馆CIP数据核字（2023）第022382号

书　　名：云计算应用运维实战
作　　者：郭立文　魏晓艳　王海龙

策　　划：翟玉峰　谢世博
责任编辑：翟玉峰　包　宁　　　　编辑部电话：（010）83525088
编辑助理：谢世博
封面设计：郑春鹏
责任校对：刘　畅
责任印制：樊启鹏

出版发行：中国铁道出版社有限公司（100054，北京市西城区右安门西街 8 号）
网　　址：http://www.tdpress.com/51eds/

印　　刷：北京联兴盛业印刷股份有限公司
版　　次：2023 年 5 月第 1 版　2023 年 5 月第 1 次印刷
开　　本：787 mm×1 092 mm　1/16　印张：13.25　字数：314 千
书　　号：ISBN 978-7-113-29943-9
定　　价：53.80 元

版权所有　侵权必究

凡购买铁道版图书，如有印制质量问题，请与本社教材图书营销部联系调换。电话：（010）63550836
打击盗版举报电话：（010）63549461

前　言

随着教学改革的深入，大部分高职院校计算机类相关专业的课程都在进行项目化教学，通过项目化教学锻炼学生综合应用和实践操作能力，以达到人才培养与企业无缝对接的目标。在此背景下，编者凭借多年教学经验，引入企业真实项目，并以教学规律、教学进程等为前提，撰写了本书，旨在为项目教学的师生提供参考。本书由具备丰富教学经验的专业课教师和企业工程师共同开发，采用企业真实项目，将工作任务转化为学习领域课程内容，学校与企业共同进行课程内容的开发与设计。教材内容组织上，摒弃了传统的知识架构，是以项目为载体，采用工作任务模式，以工作任务引领专业知识。本书适合教学做一体化教学，通过项目实战操作，培养学生职业能力和实际操作技能，使其胜任相关岗位工作。

与其他同类云计算开发实战书籍相比较，本书有以下几个特点：

（1）特别注重基础知识的讲解，适合没有云计算开发基础的学生学习。

（2）书中任务实训模块能举一反三，同时规模难度适中，适合学生课后练习。

（3）本书语言简练、通俗易懂，各任务实现给出了具体步骤，帮助学生理解所学知识。

本书包含6个项目，共18个单元，项目1包含2个单元，介绍当前流行的自动化运维工具Ansible和Ansible Playbook，内容包括Ansible的安装、Ansible Playbook的基本使用、循环、条件判断以及handlers。项目2包含5个单元，内容包括Docker容器的安装、注册Docker Hub仓库、创建Docker镜像、Docker的数据管理和数据卷容器操作以及Dockerfile使用。项目3包含3个单元，内容包括Elasticsearch工具的安装与配置、集群状态的检测、Kibana的配置、部署Logstash、使用Beats采集日志。项目4包含4个单元，内容包括MySQL密码修改、连接数据库、MySQL的常用操作、MySQL的主从配置。项目5包含2个单元，内容包括检测网站打开速度、注册和使用DNSPOD。项目6包含2个单元，内容包括JDK安装、Tomcat安装、Tomcat的配置。每个单元按照项目实现过程分解为多个任务，每个任务由任务描述、任务目标、任务实现、任务考评以及供使用者练习的实训任务构成，以完成任务的实操为主线，在任务描述中提供任务涉及的知识介绍，一些操作中应用到的知识点在任务实现的知识链接中讲解，既保证理论知识在实践操作中学习应用，也对实操提供支撑。

本书由郭立文、魏晓艳、王海龙任主编，由夏卓越、宁春梅、应天龙任副主编。

由于编者水平有限，书中疏漏及不妥之处在所难免，敬请广大读者批评指正。

编　者

2022 年 12 月

目　录

项目 1　自动化运维 1-1

单元 1　Ansible 安装与 Ansible Playbook 使用 1-2
- 任务 1　Ansible 安装 1-2
- 任务 2　Ansible Playbook 中的基本使用 1-13
- 任务 3　Ansible Playbook 中的循环 1-17
- 任务 4　Ansible Playbook 中的条件判断 1-20
- 任务 5　Ansible Playbook 中的 handlers 1-23

单元 2　Ansible Playbook 进阶使用 1-26
- 任务 1　Ansible 自动化安装 Nginx 1-26
- 任务 2　管理配置文件 1-34

项目 2　Docker 搭建与运维 2-1

单元 1　安装 Docker 与注册仓库 Docker Hub 2-2
- 任务 1　Docker 安装 2-2
- 任务 2　注册仓库 Docker Hub 2-5

单元 2　Docker 镜像管理 / 仓库管理 2-9
- 任务 1　搜索镜像与下载镜像 2-9
- 任务 2　镜像的基本操作 2-13

单元 3　通过容器创建镜像与 Docker 容器管理 2-16
- 任务 1　通过容器创建镜像 2-16
- 任务 2　Docker 容器管理 2-19

单元 4　Docker 数据管理与 Docker 网络管理 2-25
- 任务 1　Docker 数据管理之数据卷容器实践 2-25
- 任务 2　Docker 网络管理 2-28

单元 5　Dockerfile 使用 2-33
- 任务 1　自定义镜像 2-33
- 任务 2　自定义 Tomcat 9 镜像 2-38

项目 3　ELK 日志分析系统 3-1

单元 1　Elasticsearch 安装与配置 3-2
- 任务 1　基础环境配置 3-2
- 任务 2　Elasticsearch 安装 3-6
- 任务 3　配置 Elasticsearch 3-10

单元 2　检测集群状态和部署 Kibana 3-15
- 任务 1　检测集群状态 3-15

I

任务 2　部署 Kibana 3-18
　单元 3　部署 Logstash 和收集 Nginx 日志 3-21
　　任务 1　部署 Logstash 3-21
　　任务 2　Logstash 收集 Nginx 日志 3-30
　　任务 3　使用 Beats 采集日志 3-37

项目 4　MySQL 常用操作 4-1

　单元 1　更改 root 密码与连接 MySQL 4-2
　　任务 1　更改 root 密码 4-2
　　任务 2　连接 MySQL 4-5
　单元 2　MySQL 的常用操作与创建用户以及授权 4-8
　　任务 1　MySQL 的常用操作 4-8
　　任务 2　MySQL 创建用户以及授权 4-13
　单元 3　MySQL 常用 SQL 语句以及 MySQL 数据库的备份与恢复 4-16
　　任务 1　MySQL 常用 SQL 语句 4-16
　　任务 2　MySQL 数据库的备份与恢复 4-21
　单元 4　MySQL 主从配置 4-24
　　任务 1　主配置（安装完 MySQL 的虚拟机） 4-24
　　任务 2　从配置（安装完 MySQL 的虚拟机） 4-29
　　任务 3　主从同步及相关配置参数 4-33
　　任务 4　测试主从 4-36

项目 5　网站优化与服务器优化 5-1

　单元 1　了解基础知识与检测网站打开速度 5-2
　　任务 1　了解网站优化与服务器优化的基础知识 5-2
　　任务 2　检测网站打开速度 5-5
　单元 2　注册和使用 DNSPOD 与接入 CDN 厂商 5-8
　　任务 1　注册和使用 DNSPOD 5-8
　　任务 2　接入 CDN 厂商 5-12

项目 6　Tomcat 搭建配置 6-1

　单元 1　安装 JDK 与安装 Tomcat 6-2
　　任务 1　安装 JDK 6-2
　　任务 2　安装 Tomcat 6-6
　单元 2　配置 Tomcat 监听 80 端口、虚拟主机和生成日志 6-10
　　任务 1　配置 Tomcat 监听 80 端口 6-10
　　任务 2　配置 Tomcat 虚拟主机 6-15
　　任务 3　配置 Tomcat 生成日志 6-26

参考文献 C-1

资源明细表

序号	链接内容	页码
1	自动化运维及其常见工具	1-1
2	Ansible 介绍及安装	1-2
3	Ansible 远程执行命令	1-5
4	Ansible 复制文件或目录	1-5
5	Ansible 远程执行脚本	1-6
6	Ansible 管理任务计划	1-8
7	Ansible 安装 RPM 包、管理服务	1-9
8	Ansible Playbook 的使用	1-13
9	Ansible Playbook 中的循环	1-17
10	Ansible Playbook 中的条件判断	1-20
11	Ansible Playbook 中的 handlers	1-23
12	Nginx 安装 1	1-26
13	Nginx 安装 2 和 3	1-30
14	Nginx 安装 4	1-31
15	管理配置文件	1-34
16	Docker 安装	2-2
17	Docker 镜像管理、仓库管理	2-9
18	通过容器创建镜像	2-16
19	Docker 容器管理	2-19
20	Docker 数据管理	2-25
21	Docker 网络管理	2-28
22	Dockerfile 使用	2-33
23	ELK 介绍	3-1
24	基础环境配置	3-2
25	Elasticsearch 安装	3-6
26	部署 Kibana	3-18

序号	链接内容	页码
27	部署 Logstash	3-21
28	Logstash 收集 Nginx 日志	3-30
29	使用 Beats 采集日志	3-37
30	MySQL 常用操作及更改 root 密码	4-2
31	连接 MySQL	4-5
32	MySQL 常用命令	4-9
33	MySQL 创建用户以及授权	4-13
34	MySQL 常用 SQL 语句	4-16
35	MySQL 数据库的备份与恢复	4-21
36	常用网站优化的方法	5-2
37	服务器、VPS、空间的介绍	5-3
38	网站结构的演变过程	5-3
39	DNS 原理解析	5-8
40	注册和使用 DNSPOD	5-8
41	CDN 原理解析	5-12
42	接入 CDN 厂商	5-12
43	Tomcat 介绍	6-1
44	安装 JDK	6-2
45	安装 Tomcat	6-6
46	配置 Tomcat 监听 80 端口	6-10
47	配置 Tomcat 虚拟主机 1	6-15
48	配置 Tomcat 虚拟主机 2	6-17
49	配置 Tomcat 虚拟主机 3	6-22
50	配置 Tomcat 日志	6-26

项目 1　自动化运维

本项目主要介绍自动化运维的相关工具 Ansible 与 Ansible Playbook。共分为两个子单元。单元 1 介绍了 Ansible 安装与 Ansible Playbook 使用，包含 Ansible Playbook 的基本使用、循环、条件判断以及 handlers。单元 2 介绍了 Ansible Playbook 进阶使用。

视　频

自动化运维及
其常见工具

云计算应用运维实战

单元 1 Ansible 安装与 Ansible Playbook 使用

工欲善其事，必先利其器。对于工程师来说，在项目开发之前，准备工作是非常重要的环节。本单元包括两个任务：任务 1 是 Ansible 安装与 Ansible 的一些基本命令的学习，安装 Ansible 前需要先关闭防火墙和 SELinux，并修改 /etc/hosts 文件；任务 2 是 Ansible Playbook 的使用，需要学习把模块写入配置文件中与创建用户。

学习目标

通过本单元的学习，使学生掌握 Ansible Playbook 基本知识，培养学生自主进行 Ansible 安装以及常规 Ansible Playbook 使用的能力。

· 视 频
Ansible介绍及安装

任务 1 Ansible 安装

任务描述

情境描述	Ansible基于Python开发，不需要安装客户端。因为Ansible基于SSH远程管理，而Linux服务器大部分都离不开SSH，所以Ansible不需要为配置添加额外的支持。Ansible安装使用都很简单，而且基于上千个插件和模块，实现各种软件、平台、版本的管理，支持虚拟容器多层级的部署。项目主管A决定开展自动化运维的新项目，选择什么样的自动运维化工具是项目开展的重中之重。经过在网上查阅相关资料，项目主管A决定选用Ansible，它集成了众多运维工具的优点，实现了批量系统配置、批量程序部署、批量运行命令等功能
任务分解	分析上面的工作情境，将任务分解如下： （1） 环境准备； （2） 安装Ansible； （3） 免密配置； （4） 主机组设置； （5） Ansible远程执行命令； （6） Ansible复制文件或目录； （7） Ansible远程执行脚本； （8） Ansible管理任务计划； （9） Ansible安装RPM包管理服务
任务准备	了解Linux系统的基本操作，熟悉yum安装程序的过程和指令。安装Ansible是使用yum进行安装的

任务目标

知识目标	掌握Ansible软件的作用和内容
技能目标	（1）熟悉Ansible的安装； （2）了解Ansible软件的基本操作
素质目标	耐心与严谨：在下载安装Ansible过程中，解决遇到的安装路径报错、版本不符等问题，提高个人耐心与严谨的作风

任务实现

步骤1： 环境准备，在两台机器上关闭防火墙和SELinux，并修改/etc/hosts文件。

```
[root@ansible-test1 ~]# systemctl stop firewalld
[root@ansible-test1 ~]# systemctl disable firewalld
Removed symlink /etc/systemd/system/dbus-org.fedoraproject.FirewallD1.service.
Removed symlink /etc/systemd/system/basic.target.wants/firewalld.service.
[root@ansible-test1 ~]# setenforce 0
[root@ansible-test1 ~]# cat /etc/selinux/config
...
# disabled - No SELinux policy is loaded.
SELINUX=disabled                   //将此处改为disabled
# SELINUXTYPE= can take one of three two values:
...
[root@ansible-test1 ~]# cat /etc/hosts
127.0.0.1     localhost localhost.localdomain localhost4 localhost4.localdomain4
::1           localhost localhost.localdomain localhost6 localhost6.localdomain6
192.168.2.10 ansible-test1      //添加两台主机的IP和主机名
192.168.2.20 ansible-test2
```

知识链接

防火墙是指一个由软件和硬件设备组合而成，在内部网和外部网之间、专用网与公共网之间的界面上构造的保护屏障，是一种获取安全性方法的形象说法。它是一种计算机硬件和软件的结合，使Internet与Intranet之间建立起一个安全网关（Security Gateway），从而保护内部网免受非法用户的侵入。防火墙主要由服务访问规则、验证工具、包过滤和应用网关4部分组成。

步骤2： 安装Ansible，准备两台机器ansible-01和ansible-02，只需要在ansible-01上安装Ansible。先安装EPEL仓库。

```
[root@ansible-test1 ~]# yum install epel-release -y
```

```
[root@ansible-test1 ~]# yum install -y ansible
[root@ansible-test1 ~]# ansible --version
ansible 2.9.10
   config file = /etc/ansible/ansible.cfg
   configured module search path = [u'/root/.ansible/plugins/modules',
 u'/usr/share/ansible/plugins/modules']
   ansible python module location = /usr/lib/python2.7/site-packages/ansible
   executable location = /usr/bin/ansible
   python version = 2.7.5 (default, Nov 20 2015, 02:00:19) [GCC 4.8.5 20150623
(RedHat 4.8.5-4)]
```

知识链接

EPEL（Extra Packages for Enterprise Linux）由 Fedora 社区打造，为 RHEL 及衍生发行版（如 CentOS、Scientific Linux 等）提供高质量软件包的项目。安装EPEL之后，就相当于添加了一个第三方源。

步骤3： 免密配置，ansible-01上生成密钥对ssh-keygen -t rsa，把公钥放到ansible-02上，设置密钥认证。注意：本机也需要进行免密配置。

```
[root@ansible-test1 ~]# ssh-keygen -t rsa
Generating public/private rsa key pair.
Enter file in which to save the key (/root/.ssh/id_rsa):
Created directory '/root/.ssh'.
Enter passphrase (empty for no passphrase):
Enter same passphrase again:
Your identification has been saved in /root/.ssh/id_rsa.
Your public key has been saved in /root/.ssh/id_rsa.pub.
The key fingerprint is:
0a:47:86:44:83:a2:7c:c3:0c:1b:33:1c:03:88:0c:09 root@ansible-test1
The key's randomart image is:
+--[ RSA 2048]----+
|E+.o+            |
|=Bo. o           |
|o.O . o          |
|.o = o           |
| . o . S         |
|    o .          |
|     .           |
|                 |
|                 |
+-----------------+
```

```
[root@ansible-test1 ~]# ssh-copy-id 192.168.2.20
The authenticity of host '192.168.2.20 (192.168.2.20)' can't be established.
ECDSA key fingerprint is dc:a5:08:4d:9a:40:8a:be:ee:68:dd:41:61:7d:d7:05.
Are you sure you want to continue connecting (yes/no)? yes
/usr/bin/ssh-copy-id: INFO: attempting to log in with the new key(s), to
 filter out any that are already installed
/usr/bin/ssh-copy-id: INFO: 1 key(s) remain to be installed -- if you are
prompted now it is to install the new keys
root@192.168.2.20's password:
Number of key(s) added: 1

Now try logging into the machine, with:    "ssh '192.168.2.20'"
and check to make sure that only the key(s) you wanted were added.
[root@ansible-test1 ~]# ssh 192.168.2.20
Last login: Sat Jul  4 16:49:18 2020 from 192.168.2.3
[root@ansible-test2 ~]# logout
Connection to 192.168.2.20 closed.
```

步骤4： 主机组设置。在/etc/ansible/hosts文件中添加本机和另一台机器的IP：

```
[root@ansible-test1 ~]# grep ^[^#] /etc/ansible/hosts
[testhost]
127.0.0.1
192.168.2.20
```

> **说明：**
> testhost 为自定义的主机组名字；两个 IP 为组内的机器 IP。

步骤5： Ansible远程执行命令。这里的testhost为主机组名，-m后边是模块名字，-a后面是命令。当然，也可以直接写一个IP，针对某一台机器执行命令。

```
[root@ansible-test1 ~]# ansible testhost -m command -a "hostname"
127.0.0.1 | CHANGED | rc=0 >>
ansible-test1
192.168.2.20 | CHANGED | rc=0 >>
ansible-test2
[root@ansible-test1 ~]# ansible 192.168.2.20 -m command -a "hostname"
192.168.2.20 | CHANGED | rc=0 >>
ansible-test2
```

视频 Ansible远程执行命令

视频 Ansible复制文件或目录

步骤6： Ansible复制文件或目录。

源目录会放到目标目录下面，如果目标指定的目录不存在，则会自动创建。

如果复制的是文件，dest指定的名字和源不同，并且它不是已经存在的目录，则相当于复制之后又进行了重命名。如果dest是目标机器上已经存在的目录，则会直接把文件复制到该目录下面。

```
[root@ansible-test1 ~]# ansible 192.168.2.20 -m copy -a "src=/etc/passwd dest=/tmp/123"
192.168.2.20 | CHANGED => {
    "ansible_facts": {
        "discovered_interpreter_python": "/usr/bin/python"
    },
    "changed": true,
    "checksum": "8f3ebea24b1558e6207af80195aa12931d96345f",
    "dest": "/tmp/123",
    "gid": 0,
    "group": "root",
    "md5sum": "ca8f3327c9a73cb6fd96ba88ec4d18ee",
    "mode": "0644",
    "owner": "root",
    "secontext": "unconfined_u:object_r:admin_home_t:s0",
    "size": 1040,
    "src": "/root/.ansible/tmp/ansible-tmp-1593856449.24-11462-53060923085626/source",
    "state": "file",
    "uid": 0
}
```

这里的/tmp/123和源机器上的/etc/passwd是一致的。如果目标机器上已经有/tmp/123目录，则会在/tmp/123目录下面建立passwd文件。

步骤7： Ansible远程执行脚本。

（1）创建一个shell脚本。

```
[root@ansible-test1 ~]# cat /tmp/test.sh
#!/bin/bash
echo 'date' > /tmp/ansible_test.txt
```

视频
Ansible远程执行脚本

知识链接

　　shell script是利用shell的功能所写的一个程序。这个程序使用纯文本文件，写入一些shell的语法与指令，然后用正规表示法、管道命令以及数据流重导向等功能，达到所要的处理目的。

（2）把该脚本分发到各个机器上。

```
[root@ansible-test1 ~]# ansible testhost -m copy -a "src=/tmp/test.sh
dest=/tmp/test.sh
mode=0755"
    192.168.2.20 | CHANGED => {
    "ansible_facts": {
        "discovered_interpreter_python": "/usr/bin/python"
    },
    "changed": true,
    "checksum": "1a6e4af02dba1bda6fc8e23031d4447efeba0ade",
    "dest": "/tmp/test.sh",
    "gid": 0,
    "group": "root",
    "md5sum": "edfaa4371316af8c5ba354e708fe8a97",
    "mode": "0755",
    "owner": "root",
    "secontext": "unconfined_u:object_r:admin_home_t:s0",
    "size": 48,
    "src":"/root/.ansible/tmp/ansible-tmp-1593856700.7-11499-220274653312920/
source",
    "state": "file",
    "uid": 0
}
    127.0.0.1 | CHANGED => {
    "ansible_facts": {
        "discovered_interpreter_python": "/usr/bin/python"
    },
    "changed": true,
    "checksum": "1a6e4af02dba1bda6fc8e23031d4447efeba0ade",
    "dest": "/tmp/test.sh",
    "gid": 0,
    "group": "root",
    "mode": "0755",
    "owner": "root",
    "path": "/tmp/test.sh",
    "secontext": "unconfined_u:object_r:user_tmp_t:s0",
    "size": 48,
    "state": "file",
    "uid": 0
}
```

(3) 批量执行该shell脚本。

```
[root@ansible-test1 ~]# ansible testhost -m shell -a "/tmp/test.sh"
127.0.0.1 | CHANGED | rc=0 >>
192.168.2.20 | CHANGED | rc=0 >>
```

shell模块还支持远程执行命令并且带管道。

```
[root@ansible-test1 ~]# ansible testhost -m shell -a "cat /etc/passwd |wc -l "
127.0.0.1 | CHANGED | rc=0 >>
21
192.168.2.20 | CHANGED | rc=0 >>
21
[root@ansible-test1 ~]# cat /tmp/ansible_test.txt
2020年 07月 04日 星期六 18:00:51 CST
```

运行成功。

步骤8: Ansible管理任务计划。

创建任务计划,命名并定义工作。

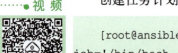

Ansible管理任务计划

```
[root@ansible-test1 ~]# ansible testhost -m cron -a "name="test cron' job='/bin/bash
/tmp/test.sh' weekday=6"
127.0.0.1 | CHANGED => {
    "ansible_facts": {
        "discovered_interpreter_python": "/usr/bin/python"
    },
    "changed": true,
    "envs": [],
    "jobs": [
        "test cron"
    ]
}
192.168.2.20 | CHANGED => {
    "ansible_facts": {
        "discovered_interpreter_python": "/usr/bin/python"
    },
    "changed": true,
    "envs": [],
    "jobs": [
        "test cron"
    ]
}
```

若要删除该cron,只需加一个字段state=absent即可。

```
[root@ansible-test1 ~]# ansible testhost -m cron -a "name='test cron' state=absent"
127.0.0.1 | CHANGED => {
    "ansible_facts": {
        "discovered_interpreter_python": "/usr/bin/python"
    },
    "changed": true,
    "envs": [],
    "jobs": []
}
192.168.2.20 | CHANGED => {
    "ansible_facts": {
        "discovered_interpreter_python": "/usr/bin/python"
    },
    "changed": true,
    "envs": [],
    "jobs": []
}
```

其他时间表示包括:分钟为minute;小时为hour;日期为day;月份为month。

步骤9: Ansible安装RPM包/管理服务。

使用yum模块安装httpd服务。

视频

Ansible安装RPM包、管理服务

```
[root@ansible-test1 ~]# ansible testhost -m yum -a "name=httpd"
127.0.0.1 | CHANGED => {
    "ansible_facts": {
        "discovered_interpreter_python": "/usr/bin/python"
    },
    "changed": true,
    "changes": {
        "installed": [
            "httpd"
        ]
    },
    "msg": "",
    "rc": 0,
    "results": [
        ...
        ...
        "\n\nComplete!\n"
    ]
}
```

```
192.168.2.20 | CHANGED => {
    "ansible_facts": {
        "discovered_interpreter_python": "/usr/bin/python"
    },
    "changed": true,
    "changes": {
        "installed": [
            "httpd"
        ]
    },
    "msg": "",
    "rc": 0,
    "results": [
        ...
        "\n\nComplete!\n"
    ]
}
```

在name后面还可以加上state=installed/removed。

设置服务状态，这里的name是CentOS系统中的服务名，可以通过chkconfig –list命令查到。

```
[root@ansible-test1 ~]# ansible testhost -m service -a "name=httpd state=started enabled=yes"
127.0.0.1 | CHANGED => {
    "ansible_facts": {
        "discovered_interpreter_python": "/usr/bin/python"
    },
    "changed": true,
    "enabled": true,
    "name": "httpd",
    "state": "started",
    "status": {
        ...
        "WatchdogTimestampMonotonic": "0",
        "WatchdogUSec": "0"
    }
}
192.168.2.20 | CHANGED => {
    "ansible_facts": {
        "discovered_interpreter_python": "/usr/bin/python"
    },
    "changed": true,
    "enabled": true,
```

```
    "name": "httpd",
    "state": "started",
    "status": {
      ...
        "WatchdogUSec": "0"
    }
}
```

Ansible文档的使用见图1-1-1和图1-1-2。

```
[root@ansible-test1 ~]# ansible-doc -l
fortios_router_community_list                      Configure community lists in Fortin...
azure_rm_devtestlab_info                           Get Azure DevTest Lab facts
ecs_taskdefinition                                 register a task definition in ecs
avi_alertscriptconfig                              Module for setup of AlertScriptConf...
tower_receive                                      Receive assets from Ansible Tower
netapp_e_iscsi_target                              NetApp E-Series manage iSCSI target...
azure_rm_acs                                       Manage an Azure Container Service(A...
fortios_log_syslogd2_filter                        Filters for remote system server in...
junos_rpc                                          Runs an arbitrary RPC over NetConf ...
na_elementsw_vlan                                  NetApp Element Software Manage VLAN
pn_ospf                                            CLI command to add/remove ospf prot...
pn_snmp_vacm                                       CLI command to create/modify/delete...
cp_mgmt_service_sctp                               Manages service-sctp objects on Che...
onyx_ospf                                          Manage OSPF protocol on Mellanox ON...
icx_command                                        Run arbitrary commands on remote Ru...
cs_snapshot_policy                                 Manages volume snapshot policies on...
nxos_install_os                                    Set boot options like boot, kicksta...
cnos_static_route                                  Manage static IP routes on Lenovo C...
win_eventlog                                       Manage Windows event logs
vmware_category                                    Manage VMware categories
vmware_host_feature_info                           Gathers info about an ESXi host's f...
avi_cluster                                        Module for setup of Cluster Avi RES...
na_ontap_user                                      NetApp ONTAP user configuration and...
aci_l3out                                          Manage Layer 3 Outside (L3Out) obje...
memset_server_info                                 Retrieve server information
gcp_compute_subnetwork_info                        Gather info for GCP Subnetwork
azure_rm_virtualmachinescalesetextension           Manage Azure Virtual Machine Scale ...
fortios_report_dataset                             Report dataset configuration in For...
avi_api_session                                    Avi API Module
avi_networkprofile                                 Module for setup of NetworkProfile ...
avi_backup                                         Module for setup of Backup Avi REST...
aci_interface_policy_cdp                           Manage CDP interface policies (cdp:...
fortios_firewall_vip                               Configure virtual IP for IPv4 in Fo...
gcp_compute_backend_service                        Creates a GCP BackendService
iam_policy                                         Manage IAM policies for users, grou...
fortios_system_fips_cc                             Configure FIPS-CC mode in Fortinet'...
fortios_log_null_device_setting                    Settings for null device logging in...
win_inet_proxy                                     Manages proxy settings for WinINet ...
fortios_firewall_auth_portal                       Configure firewall authentication p...
jenkins_plugin                                     Add or remove Jenkins plugin
```

图 1-1-1　Ansible 的使用之列出所有模块

```
[root@ansible-test1 ~]# ansible-doc yum
> YUM    (/usr/lib/python2.7/site-packages/ansible/modules/packaging/os/yum.py)

        Installs, upgrade, downgrades, removes, and lists packages and groups with the `yum'
        package manager. This module only works on Python 2. If you require Python 3 support
        see the [dnf] module.

  * This module is maintained by The Ansible Core Team
  * note: This module has a corresponding action plugin.

OPTIONS (= is mandatory):

- allow_downgrade
        Specify if the named package and version is allowed to downgrade a maybe already
        installed higher version of that package. Note that setting allow_downgrade=True can
        make this module behave in a non-idempotent way. The task could end up with a set of
        packages that does not match the complete list of specified packages to install
        (because dependencies between the downgraded package and others can cause changes to
        the packages which were in the earlier transaction).
        [Default: no]
        type: bool
        version_added: 2.4

- autoremove
        If `yes', removes all "leaf" packages from the system that were originally installed
        as dependencies of user-installed packages but which are no longer required by any
        such package. Should be used alone or when state is `absent'
        NOTE: This feature requires yum >= 3.4.3 (RHEL/CentOS 7+)
        [Default: no]
        type: bool
        version_added: 2.7

- bugfix
        If set to `yes', and `state=latest' then only installs updates that have been marked
        bugfix related.
        [Default: no]
        version_added: 2.6

- conf_file
        The remote yum configuration file to use for the transaction.
        [Default: (null)]
        version_added: 0.6

- disable_excludes
        Disable the excludes defined in YUM config files.
        If set to `all', disables all excludes.
        If set to `main', disable excludes defined in [main] in yum.conf.
```

图 1-1-2　Ansible 的使用之查看指定模块的文档

任务考评

考评记录

姓名			完成日期	
序号	考核内容		标准分	评分
1	Ansible 的 yum 安装		10	
2	免密配置		10	
3	主机组设置		10	
4	Ansible 远程执行命令		10	
5	Ansible 复制文件或目录		20	
6	Ansible 远程执行脚本		20	
7	Ansible 管理任务计划		20	
	总评分		100	

任务实现心得：

任务实训

任务实训	远程执行命令时尝试执行不同的命令
任务目标	学会 Ansible 执行命令的方式

任务 2 Ansible Playbook 中的基本使用

任务描述

情境描述	程序员A在查找Ansible相关资料时发现了Ansible Playbook工具。Playbook是一个不同于使用Ansible命令行执行方式的模式，其功能更加强大。简单来说，Playbook是一个非常简单的配置管理和多主机部署系统，可作为一个适合部署复杂应用程序的基础。Playbook可以定制配置，可以按照指定的操作步骤有序执行，支持同步和异步方式。所以，A决定认真学习一下
任务分解	分析上面的工作情境，将任务分解如下： (1) 模块写入配置文件； (2) 创建用户的例子
任务准备	安装Ansible软件，熟悉Linux基本操作

任务目标

知识目标	了解Playbook的基本使用
技能目标	配置文件中写入模块，学会创建用户的例子
素质目标	耐心与严谨：在Playbook的使用过程中，解决遇到的各种文件配置问题，提高个人的耐心与严谨的作风

任务实现

步骤1： 把模块写入配置文件中。

```
[root@ansible-test1 ansible]# cat /etc/ansible/test.yml
---
- hosts: 192.168.2.20
  remote_user: root
  tasks:
  - name: test_playbook
    shell: touch /tmp/playbook_test.txt
```

视频

Ansible Playbook的使用

说明： 第一行后需要有三个杠；hosts参数指定了对哪些主机进行操作，如果是多台机器可以用逗号作为分隔，也可以使用主机组，在 /etc/ansible/hosts 中定义；user 参数指定了使用什么用户登录远程主机操作；tasks 指定了一个任务，其下的 name 参数是对任务的描述，在执行过程中会打印出来；shell 是 Ansible 模块的名字。

```
[root@ansible-test1 ansible]# ansible-playbook test.yml
PLAY[192.168.2.20]
***************************************************************************
TASK[GatheringFacts]
***************************************************************************
ok: [192.168.2.20]
TASK[test_playbook]
***************************************************************************
[WARNING]: Consider using the file module with state=touch rather than running
'touch'.  If you need to use command because file is insufficient you can add 'warn:
false' to this command task or set
'command_warnings=False' in ansible.cfg to get rid of this message.
changed: [192.168.2.20]
PLAYRECAP
***************************************************************************
192.168.2.20    :ok=2   changed=1 unreachable=0 failed=0 skipped=0 rescued=0 ignored=0
```

步骤2： 创建用户。

```
[root@ansible-test1 ansible]# cat create_user.yml
---
- name: create_user
  hosts: 192.168.2.20
  user: root
  gather_facts: false
  vars:
    - user: "test"
  tasks:
    - name: create user
      user: name="{{ user }}"
```

> **说明：**
>
> name 参数对该 Playbook 实现的功能做一个概述，后面执行过程中，会打印 name 变量的值，可以省略；gather_facts 参数指定了在以下任务部分执行前是否先执行 setup 模块获取主机相关信息，这在后面的 task 使用到 setup 获取的信息时用到；vars 参数指定了变量，这里指定一个 user 变量，其值为 test，需要注意的是，变量值一定要用引号引住；user 参数指定了调用 user 模块，name 是 user 模块中的一个参数，而增加的用户名字调用了上面 user 变量的值。

```
[root@ansible-test1 ansible]# ansible-playbook create_user.yml
PLAY[create_user]
***************************************************************************
TASK[createuser]
```

```
**************************************************************************
changed: [192.168.2.20]
PLAYRECAP
**************************************************************************
192.168.2.20              : ok=1    changed=1   unreachable=0   failed=0
skipped=0    rescued=0    ignored=0
```

学习笔记

考评记录

姓名			完成日期	
序号	考核内容		标准分	评分
1	模块写入配置文件		50	
2	创建用户		50	
	总评分		100	

任务实现心得：

任务实训	修改配置文件中的参数，查看会有什么结果
任务目标	掌握配置文件中各个参数的含义

任务 3　Ansible Playbook 中的循环

任务描述

情境描述	程序员C认为循环是程序编写中必不可少的部分，所以他想研究一下如何在Ansible Playbook中实现循环
任务分解	分析上面的工作情境，将任务分解如下： （1）创建while.yml文件； （2）执行while.yml文件
任务准备	了解Playbook的基本使用方式

任务目标

知识目标	掌握循环的基本实现
技能目标	（1）学会创建循环文件； （2）学会执行文件的方式
素质目标	耐心与严谨：在Playbook循环使用过程中，解决遇到的各种循环代码编写和执行问题，提高个人的耐心与严谨的作风

任务实现

步骤1： 创建while.yml文件。

```
[root@ansible-test1 ansible]# cat while.yml
---
- hosts: testhost
  user: root
  tasks:
    - name: change mode for files
      file: path=/tmp/{{ item }} mode=600
      with_items:
        - 1.txt
        - 2.txt
        - 3.txt
```

视频

Ansible Playbook中的循环

> **说明：**
> with_items 为循环的对象。

步骤2： 执行while.yml。

```
[root@ansible-test1 ansible]# ansible-playbook while.yml
PLAY[testhost]
************************************************************************
TASK[Gathering Facts]
************************************************************************
ok: [127.0.0.1]
ok: [192.168.2.20]
TASK[change mode forfiles]
************************************************************************
ok: [127.0.0.1] => (item=1.txt)
changed: [192.168.2.20] => (item=1.txt)
ok: [127.0.0.1] => (item=2.txt)
changed: [192.168.2.20] => (item=2.txt)
ok: [127.0.0.1] => (item=3.txt)
changed: [192.168.2.20] => (item=3.txt)
PLAY                                                                RECAP
************************************************************************
127.0.0.1    : ok=2 changed=0 unreachable=0 failed=0 skipped=0 rescued=0 ignored=0
192.168.2.20 : ok=2 changed=1 unreachable=0 failed=0 skipped=0 rescued=0 ignored=0
```

学习笔记

考评记录

姓名		完成日期	
序号	考核内容	标准分	评分
1	创建 while.yml 文件	50	
2	执行 while.yml 文件	50	
	总评分	100	

任务实现心得：

任务实训

任务实训	仿照并修改循环文件内容，使之正常执行
任务目标	学会循环的语法格式

任务 4　Ansible Playbook 中的条件判断

任务描述

情境描述	程序员B看到了之前程序员写的循环程序，想尝试一下Ansible Playbook中的条件判断
任务分解	分析上面的工作情境，将任务分解如下： （1）创建when.yml文件； （2）执行when.yml文件
任务准备	了解Playbook的基本使用方式

任务目标

知识目标	掌握条件判断程序
技能目标	分析上面的工作情境，将任务分解如下： （1）学会创建循环文件； （2）掌握执行文件的方式
素质目标	耐心与严谨：在Playbook条件判断使用过程中，解决遇到的各种条件判断代码编写和执行问题，培养耐心与严谨的作风

任务实现

视频

Ansible Playbook中的条件判断

步骤1： 创建when.yml文件。

```
[root@ansible-test1 ansible]# cat when.yml
---
- hosts: testhost
  user: root
  gather_facts: True
  tasks:
    - name: use when
      shell: touch /tmp/when.txt
      when: ansible_eno16777736.ipv4.address == "192.168.2.20"
```

> **说明：**
> 使用 ansible ansible-02 -m setup 命令可以查看到所有 facter 信息。

1-20

> **知识链接**
>
> YML文件格式是YAML（YAML Ain't Markup Language）编写的文件格式，YAML是一种能够被计算机识别的直观的数据序列化格式，并且容易被人类阅读，容易和脚本语言交互，可以被支持YAML库的不同编程语言程序导入，如C/C++、Ruby、Python、Java、Perl、C#、PHP等。

步骤2： 执行when.yml文件。

```
[root@ansible-test1 ansible]# ansible-playbook when.yml
PLAY [testhost]
***************************************************************************
TASK [Gathering Facts]
***************************************************************************
ok: [127.0.0.1]
ok: [192.168.2.20]

TASK [use when]
***************************************************************************
skipping: [127.0.0.1]
[WARNING]: Consider using the file module with state=touch rather than running
'touch'.  If you need to use command because file is insufficient you can add 'warn:
false' to this command task or set'command_warnings=False' in ansible.cfg to get
rid of this message.
changed: [192.168.2.20]

PLAY RECAP
***************************************************************************
127.0.0.1      : ok=1  changed=0  unreachable=0  failed=0  skipped=1  rescued=0    ignored=0
192.168.2.20   : ok=2  changed=1  unreachable=0  failed=0  skipped=0    rescued=0
ignored=0
```

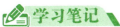

考评记录

姓名			完成日期	
序号	考核内容		标准分	评分
1	创建 while.yml 文件		50	
2	执行 while.yml 文件		50	
	总评分		100	

任务实现心得：

任务实训

任务实训	仿照并修改循环文件内容，使之正常执行
任务目标	掌握条件判断语法格式

任务 5 Ansible Playbook 中的 handlers

任务描述

情境描述	程序员A发现程序员B修改了某些程序的配置文件以后，有可能需要重启应用程序，以便使新的配置生效，那么，如何使用Playbook来实现这个功能呢？如果配置文件发生了改变，则重启服务；如果配置文件并没有被真正修改，则不对服务进行任何操作。该如何实现上述功能呢？handlers可以解决这些问题
任务分解	分析上面的工作情境，将任务分解如下：创建handlers.yml文件
任务准备	了解Playbook的基本使用方式

任务目标

知识目标	了解handlers的意义和用法
技能目标	能够配置handlers文件
素质目标	细心与耐心：在了解handlers意义和用法的过程中，耐心查看相关资料；在handlers配置文件编写过程中，细心查看配置信息，并查看handlers的工作结果

任务实现

步骤： 创建handlers.yml。

执行task之后，服务器发生变化后可能要执行一些操作，如修改了配置文件后需要重启服务，创建handlers.yml文件，加入如下内容：

视频

Ansible Playbook中的handlers

```
[root@ansible-test1 ansible]# cat handlers.yml
---
- name: handlers test
  hosts: 192.168.2.20
  user: root
  tasks:
    - name: copy file
      copy: src=/etc/passwd dest=/tmp/aaa.txt
      notify: test handlers
  handlers:
    - name: test handlers
      shell: echo "111111" >> /tmp/aaa.txt
```

> **说明：**
> 只有 copy 模块真正执行后，才会去调用下面的 handlers 相关操作。也就是说，如果源文件和目标文件内容是一样的，并不会去执行 handlers 中的 shell 相关命令。其比较适合配置文件发生更改后重启服务的操作。

```
[root@ansible-test1 ansible]# ansible-playbook handlers.yml
PLAY [handlers test]
*******************************************************************************
TASK [Gathering Facts]
*******************************************************************************
ok: [192.168.2.20]
TASK [copy file]
*******************************************************************************
changed: [192.168.2.20]
RUNNING HANDLER [test handlers]
*******************************************************************************
changed: [192.168.2.20]
PLAY RECAP
*******************************************************************************
192.168.2.20    : ok=3    changed=2    unreachable=0    failed=0    skipped=0    rescued=0    ignored=0
```

学习笔记

 **任务考评**

考评记录

姓名		完成日期	
序号	考核内容	标准分	评分
1	创建 handlers.yml 文件	50	
2	实现 handlers 功能	50	
	总评分	100	

任务实现心得：

 任务实训

任务实训	修改配置文件，观察是否重启服务
任务目标	掌握 handlers 的工作原理

云计算应用运维实战

单元 2　Ansible Playbook 进阶使用

本单元主要学习 Ansible Playbook 进阶使用，主要包括 Ansible 自动化安装 Nginx 以及管理配置文件，创建 Django 工程文件。

学习目标

通过本单元的学习，使学生掌握 Ansible Playbook 进阶知识，培养学生增加 IT 运维质量、降低成本的能力。

任务 1　Ansible 自动化安装 Nginx

任务描述

情境描述	程序员A使用源码包远程编译安装Nginx，而且通过Playbook发送到远程主机。需要远程主机自动安装Nginx，就需要使用到Ansible工具
任务分解	分析以上工作情境，将任务分解如下： （1）编译安装Nginx； （2）环境准备； （3）执行文件
任务准备	熟练掌握Ansible软件，了解Nginx的作用

任务目标

知识目标	使用Ansible自动化安装Nginx
技能目标	学会编译安装Nginx的方法
素质目标	严谨精神：在实际项目中保持严谨的态度，在Nginx的安装过程中每一步骤都确保正确

任务实现

● 视频

Nginx安装1

步骤1： 编译安装Nginx。

（1）使用wget下载Nginx包。

（2）解压下载的Nginx包。

```
./configure --prefix=/usr/local/nginx
make && make install
```

1-26

(3) 编写/etc/init.d/nginx文件。
内容如下：

```bash
#!/bin/bash
# chkconfig: - 30 21
# description: http service.
# Source Function Library
. /etc/init.d/functions
# Nginx Settings
NGINX_SBIN="/usr/local/nginx/sbin/nginx"
NGINX_CONF="/usr/local/nginx/conf/nginx.conf"
NGINX_PID="/usx/local/nginx/logs/nginx.pid"
RETVAL=0
prog="Nginx"

start()
{
    echo -n $"Starting $prog: "
    mkdir -p /dev/shm/nginx_temp
    daemon $NGINX_SBIN -c $NGINX_CONF
    RETVAL=$?
    echo
    return $RETVAL
}
stop()
{
    echo -n $"Stopping $prog: "
    killproc -p $NGINX_PID $NGINX_SBIN -TERM
    rm -rf /dev/shm/nginx_temp
    RETVAL=$?
    echo
    return $RETVAL
}
reload()
{
    echo -n $"Reloading $prog: "
    killproc -p $NGINX_PID $NGINX_SBIN -HUP
    RETVAL=$?
    echo
    return $RETVAL
}
restart()
{
    stop
    start
}
configtest()
{
```

```
            $NGINX_SBIN -c $NGINX_CONF -t
            return 0
    }
    case "$1" in
        start)
            start
            ;;
        stop)
            stop
            ;;
        reload)
            reload
            ;;
        restart)
            restart
            ;;
        configtest)
            configtest
            ;;
        *)
            echo $"Usage: $0 {start|stop|reload|restart|configtest}"
            RETVAL=1
    esac
    exit $RETVAL
```

(4) 清空配置文件并重新编写。

```
# > /usr/local/nginx/conf/nginx.conf
```

内容如下：

```
user nobody nobody;                    //定义Nginx运行的用户和用户组
worker_processes 2;                    //Nginx进程数，一般为CPU总核心数
error_log /usr/local/nginx/logs/nginx_error.log crit;    //全局错误日志定义类型
pid /usr/local/nginx/logs/nginx.pid;                     //进程文件
worker_rlimit_nofile 51200;
events                                 //工作模式与连接数上限
{
    use epoll;
    worker_connections 6000;
}
http                                   //http下的一些配置
{
    include mime.types;                //文件扩展名与文件类型映射表
    default_type application/octet-stream;              //默认文件类型
    server_names_hash_bucket_size 3526;
```

```
server_names_hash_max_size 4096;
log_format combined_realip '$remote_addr $http_x_forwarded_for [$time_local]'
'$host "$request_uri" $status'
'"$http_referer" "$http_user_agent"';
sendfile on;                    //开启高效文件传输模式
tcp_nopush on;                  //防止网络阻塞
keepalive_timeout 30;           //长连接超时时间，单位为秒
client_header_timeout 3m;
client_body_timeout 3m;
send_timeout 3m;
connection_pool_size 256;
client_header_buffer_size 1k;
large_client_header_buffers 8 4k;
request_pool_size 4k;
output_buffers 4 32k;
postpone_output 1460;
client_max_body_size 10m;
client_body_buffer_size 256k;
client_body_temp_path /usr/local/nginx/client_body_temp;
proxy_temp_path /usr/local/nginx/proxy_temp;
fastcgi_temp_path /usr/local/nginx/fastcgi_temp;
fastcgi_intercept_errors on;
tcp_nodelay on;                 //防止网络阻塞
gzip on;                        //开启gzip压缩输出
gzip_min_length 1k;
gzip_buffers 4 8k;
gzip_comp_level 5;
gzip_http_version 1.1;
gzip_types text/plain application/x-javascript text/css text/htm
application/xml;
server                          //虚拟主机配置
{
    listen 80;
    server_name localhost;
    index index.html index.htm index.php;
    root /usr/local/nginx/html;
    location ~ \.php$
    {
        include fastcgi_params;
        fastcgi_pass unix:/tmp/php-fcgi.sock;
        fastcgi_index index.php;
        fastcgi_param SCRIPT_FILENAME /usr/local/nginx/html$fastcgi_script_name;
    }
}
```

（5）编写完成后进行检查。

/usr/local/nginx/sbin/nginx -t

（6）启动Nginx。

servicenginxstart

编译安装完成。

步骤2： 环境准备。

Nginx安装2和3

（1）将nginx.tar.gz复制到/etc/ansible/nginx_install/roles/install/files下。启动脚本和配置文件都放到/etc/ansible/nginx_install/roles/install/template下。

```
# mv nginx.tar.gz /etc/ansible/nginx_install/roles/install/files/
# cp nginx/conf/nginx.conf /etc/ansible/nginx_install/roles/install/templates/
# cp /etc/init.d/nginx /etc/ansible/nginx_install/roles/install/templates/
```

（2）编写需要的yml文件。

```
[root@ansible2 nginx_install]# cat install.yml
---
- hosts: 192.168.2.101             //入口文件
  remote_user: root
  gather_facts: True
  roles:
    - common
    - install
[root@ansible2 nginx_install]# cat roles/common/tasks/main.yml
- name: install initialization require software    //安装需要的依赖
  yum: name={{ item }} state=installed
  with_items:
    - zlib-devel
    - pcre-devel
    - gcc
[root@ansible2 nginx_install]# cat roles/install/vars/main.yml
nginx_user: www                             //定义所需变量
nginx_port: 80
nginx_basedir: /usr/local/nginx
[root@ansible2 nginx_install]# cat roles/install/tasks/copy.yml
- name: Copy Nginx Software              //复制压缩包
  copy: src=nginx.tar.gz dest=/tmp/nginx.tar.gz owner=root group=root
- name: Uncompression Nginx Software     //解压压缩包
  shell: tar zxf /tmp/nginx.tar.gz -C /usr/local/
- name: Copy Nginx Start Script          //复制启动脚本
  template: src=nginx dest=/etc/init.d/nginx owner=root group=root mode=0755
```

```
    - name: Copy Nginx Config                    //复制Nginx配置文件
      template: src=nginx.conf dest={{ nginx_basedir }}/conf/ owner=root group=root
    mode=0644
    [root@ansible2 nginx_install]# cat roles/install/tasks/install.yml
    - name: create nginx user                    //创建用户
      user: name={{ nginx_user }} state=present createhome=no shell=/sbin/nologin
    - name: start nginx service                  //开启服务
      shell: /etc/init.d/nginx start
    - name: add boot start nginx service         //加入开机启动
      shell: chkconfig --level 345 nginx on
    - name: delete nginx compression files       //删除压缩包
      shell: rm -rf /tmp/nginx.tar.gz
    [root@ansible2 nginx_install]# cat roles/install/tasks/main.yml
    - include: copy.yml                          //调用copy.yml和install.yml
    - include: install.yml
```

步骤3: 执行文件。

运行install.yml文件。

```
# ansible-playbook /etc/ansible/nginx_install/install.yml
```

Nginx安装4

注意:
要检查远程机器是否存在端口占用，并及时卸载。

结果如下：

```
[root@ansible-01 ~]# ansible-playbook /etc/ansible/nginx_install/install.yml

PLAY [192.168.2.31]
************************************************************
TASK [Gathering Facts]
************************************************************
ok: [192.168.2.31]
TASK [common : install initializtion requre software]
************************************************************
 [DEPRECATION WARNING]: Invoking "yum" only once while using a loop via squash_actions
is deprecated. Instead of using a loop to supply multiple items and specifying 'name:
"{{ item }}"', please use `name: ['zlib-devel', 'pcre-devel']` and remove the loop. This
feature will be removed in version 2.11. Deprecation warnings can be disabled by setting
deprecation_warnings=False in ansible.cfg.
ok: [192.168.2.31] => (item=[u'zlib-devel', u'pcre-devel'])

TASK [install : Copy Nginx Software]
************************************************************
changed: [192.168.2.31]
```

1-31

```
TASK [install : Uncompression Nginx Software]
************************************************************************
 [WARNING]: Consider using the unarchive module rather than running 'tar'.  If
you need to use command because unarchive is insufficient you can add 'warn: false'
to this command task or set 'command_warnings=False' in ansible.cfg to get
 rid of this message.
 changed: [192.168.2.31]

TASK [install : Copy Nginx Start Script]
************************************************************************
ok: [192.168.2.31]

TASK [install : Copy Nginx Config]
************************************************************************
ok: [192.168.2.31]

TASK [install : Create Nginx User]
************************************************************************
ok: [192.168.2.31]

TASK [install : Start Nginx Service]
************************************************************************
changed: [192.168.2.31]

TASK [install : Add Boot start Nginx service]
************************************************************************
changed: [192.168.2.31]

TASK [install : Delete Nginx compression files]
************************************************************************
 [WARNING]: Consider using the file module with state=absent rather than running 'rm'.
If you need to use command because file is insufficient you can add 'warn: false'
to this command task or set 'command_warnings=False' in ansible.cfg to get rid
of this message.
changed: [192.168.2.31]

PLAY RECAP
************************************************************************
  192.168.2.31   : ok=10   changed=5   unreachable=0   failed=0   skipped=0   rescued=0
ignored=0
```

考评记录

姓名		完成日期	
序号	考核内容	标准分	评分
1	编译安装 Nginx	30	
2	环境准备	30	
3	执行文件	40	
	总评分	100	

任务实现心得：

任务实训

任务实训	修改配置文件中的信息，并观察不同
任务目标	掌握配置文件中各参数的意义

任务 2　管理配置文件

任务描述

情境描述	程序员C发现随着安装程序数越来越多，配置文件也越来越多，难以管理。所以有必要写个管理Nginx配置文件的Playbook
任务分解	管理Nginx配置文件的Playbook的编写
任务准备	分析以上工作情境，将任务分解如下： （1）了解Nginx配置文件的含义； （2）了解Playbook的基本操作

任务目标

知识目标	掌握配置文件的管理
技能目标	编写Playbook程序
素质目标	严谨精神：在实际项目中保持严谨的态度，在Playbook的设置过程中每一步骤都确保正确

视　频

管理配置文件

任务实现

步骤：管理Nginx配置文件的Playbook编写。

```
[root@ansible2 nginx_config]# cat
/etc/ansible/nginx_config/roles/new/handlers/main.yml
 - name: restart nginx               //用于重新加载Nginx服务
   shell: /etc/init.d/nginx reload
[root@ansible2 nginx_config]# cat /etc/ansible/nginx_config/roles/new/tasks/main.yml
 - name: copy conf file              //复制.conf和hosts文件
   copy: src={{ item.src }} dest={{ nginx_basedir }}/{{ item.dest }} backup=yes
owner=root group=root mode=0644
   with_items:
     - { src: nginx.conf, dest: conf/nginx.conf }
     - { src: vhosts, dest: conf/ }
   notify: restart nginx
[root@ansible2 nginx_config]# cat /etc/ansible/nginx_config/roles/new/vars/main.yml
nginx_basedir: /usr/local/nginx       //定义变量
[root@ansible2 nginx_config]# cat update.yml
---
 - hosts: 192.168.2.101                //入口文件
   user: root
```

```
  roles:
  - new              //这里只有new
```

old目录中的yml文件与new目录中的相同，files中的配置文件不同。

其中new为更新时用到的，old为回滚时用到的，files下面为Nginx.conf和vhosts目录，handlers为重启Nginx服务的命令。

在执行update.yml前，应备份当前配置文件，当执行之后发现错误，则进行回滚操作。关于回滚，需要在执行Playbook之前先备份一下旧的配置，所以对于老配置文件的管理一定要严格，千万不能随便去修改线上机器的配置，并且要保证new/files下面的配置和线上的配置一致，命令如下：

```
# rsync -av /etc/ansible/nginx_config/roles/new/
/etc/ansible/nginx_config/roles/old/
```

回滚操作就是把旧的配置覆盖，然后重新加载Nginx服务，每次改动Nginx配置文件之前先备份到old中，对应目录为/etc/ansible/nginx_config/roles/old/files。

更新操作结果：

```
[root@ansible-01 nginx_config]# ansible-playbook /etc/ansible/nginx_config/update.yml

PLAY [testhost] ****************************************************************

TASK [Gathering Facts] *********************************************************
ok: [192.168.2.31]
ok: [127.0.0.1]
TASK [new : copy conf file] ****************************************************
ok: [192.168.2.31] => (item={u'dest': u'conf/nginx.conf', u'src': u'nginx.conf'})
ok: [127.0.0.1] => (item={u'dest': u'conf/nginx.conf', u'src': u'nginx.conf'})
ok: [127.0.0.1] => (item={'dest': u'conf/', u'src': u'vhosts'})
changed: [192.168.2.31] => (item={u'dest': u'conf/', u'src': u'vhosts'})

RUNNING HANDLER [new : restart nginx] ******************************************
changed: [192.168.2.31]


PLAY RECAP *********************************************************************
127.0.0.1     : ok=2  changed=0  unreachable=0  failed=0  skipped=0  rescued=0  ignored=0
192.168.2.31  : ok=3  changed=2  unreachable=0  failed=0  skipped=0  rescued=0  ignored=0
```

回滚操作结果：

```
[root@ansible-01 nginx_config]# ansible-playbook /etc/ansible/nginx_config/rollback.yml

PLAY [testhost] *************************************************************

TASK [Gathering Facts] ******************************************************
ok: [192.168.2.31]
ok: [127.0.0.1]

TASK [old : copy conf file] *************************************************
ok: [192.168.2.31] => (item={u'dest': u'conf/nginx.conf', u'src': u'nginx.conf'})
ok: [127.0.0.1] => (item={u'dest': u'conf/nginx.conf', u'src': u'nginx.conf'})
ok: [192.168.2.31] => (item={u'dest': u'conf/', u'src': u'vhosts'})
ok: [127.0.0.1] => (item={u'dest': u'conf/', u'src': u'vhosts'})

PLAY RECAP ******************************************************************
127.0.0.1        : ok=2  changed=0  unreachable=0  failed=0  skipped=0  rescued=0  ignored=0
192.168.2.31     : ok=2  changed=0  unreachable=0  failed=0  skipped=0  rescued=0  ignored=0
```

> **注意：**
> 本次任务并未改变配置文件，故 changed 为 0。

学习笔记

 **任务考评**

考评记录

姓名		完成日期	
序号	考核内容	标准分	评分
1	管理配置文件的 Playbook 编写	50	
2	测试通过	50	
	总评分	100	
任务实现心得：			

 任务实训

任务实训	修改 Playbook 管理 Nginx 的配置文件
任务目标	掌握 Palybook 的管理方式

学习笔记

项目 2
Docker 搭建与运维

本项目主要介绍 Docker 容器工具。Docker 是一个开源的应用容器引擎，让开发者可以打包其应用以及依赖包到一个可移植的镜像中，然后发布到安装有 Linux 或 Windows 操作系统的机器上，也可以实现虚拟化。本项目共分为 5 个单元，分别介绍 Docker 的安装、镜像管理/仓库管理、通过容器创建镜像、Docker 数据管理与网络管理、Dockerfile 使用。

单元 1　安装 Docker 与注册仓库 Docker Hub

本单元包括两个任务：任务 1 是安装 Docker，安装 Docker 包括安装必要的依赖包、添加软件源信息、更新并安装 Docker-CE、开启 Docker 服务任务；任务 2 是注册仓库 Docker Hub，需要注册账号，登录邮箱激活以及登录 Docker Hub。

学习目标

通过本单元的学习，使学生掌握安装 Docker 与注册仓库 Docker Hub 的基本知识，培养学生自主进行安装 Docker 与注册仓库 Docker Hub 的能力。

任务 1　Docker 安装

视频

Docker安装

任务描述

情境描述	Docker 是一个开源的应用容器引擎，它完全使用沙箱机制，相互之间不会有任何接口（类似 iPhone 的 App），几乎没有性能开销，可以很容易地在机器和数据中心中运行。最重要的是，它们不依赖任何语言、框架或者系统。作为一种新兴的虚拟化方式，Docker 和传统的虚拟化方式相比具有众多的优势。自 2013 年推出以来，Docker 受到越来越多开发者的关注。不管是云服务还是微服务（Microservices），越来越多的厂商都开始基于 Docker 作为基础设施自动化的工具。 Docker 主要解决的问题： （1）保证程序运行环境的一致性； （2）降低配置开发环境、生产环境的复杂度及成本； （3）实现程序的快速部署和分发。 程序员 A 接到了项目经理的工作安排，为目前的项目环境搭建 Docker，以保证后期移植顺利，所以接下来是安装部分。
任务分解	分析上面的工作情境，将任务分解如下： （1）安装必要的依赖包； （2）添加软件源信息； （3）更新并安装 Docker-CE； （4）开启 Docker 服务。

任务准备	（1）在 CentOS 7 机器上关闭防火墙和 SELinux；
	（2）关闭 firewalld；
	
	`[root@docker ~]# systemctl stop firewalld`
	`[root@docker ~]# systemctl disable firewalld`

任务准备	(3) 关闭SElinux：
	`[root@docker ~]# setenforce 0`
	`[root@docker ~]# sed -i 's/SELINUX=enforcing/SELINUX=disabled/g' /etc/selinux/config`

任务目标

知识目标	了解Docker并掌握Docker安装方法
技能目标	能够独立正确安装Docker
素质目标	耐心与严谨：在安装Docker的过程中，解决诸如软件源信息错误、依赖包版本错误等问题，提高个人的耐心与严谨的作风

任务实现

步骤1： 安装必要的依赖包。

```
[root@docker ~]# yum install -y yum-utils device-mapper-persistent-data lvm2
```

步骤2： 添加软件源信息。

```
[root@docker ~]# yum-config-manager --add-repohttp://mirrors.aliyun.com/docker-ce/linux/centos/docker-ce.repo
```

步骤3： 更新并安装Docker-CE。

```
[root@docker ~]# yum makecache fast
[root@docker ~]# yum -y install docker-ce
```

步骤4： 开启Docker服务。

```
[root@docker ~]# systemctl daemon-reload
[root@docker ~]# systemctl restart docker
[root@docker ~]# systemctl enable docker
Created symlink from /etc/systemd/system/multi-user.target.wants/docker.service to /usr/lib/systemd/system/docker.service.
```

任务考评

考评记录

姓名			完成日期	
序号	考核内容		标准分	评分
1	成功安装必要的依赖包		25	
2	成功添加软件源信息		25	
3	成功更新并安装 Docker-CE		25	
4	成功开启 Docker 服务		25	
	总评分		100	

任务实现心得：

任务实训

任务实训	实现 Docker 安装的全过程
任务目标	掌握 Docker 安装的要领

2-4

任务 2　注册仓库 Docker Hub

📖 任务描述

情境描述	Docker镜像是一个不包含Linux内核而又精简的Linux操作系统。在项目经理的安排下，大家已经安装了Docker，程序员A想要使用它，首先就要注册仓库Docker Hub
任务分解	分析上面的工作情境，将任务分解如下： (1) 登录Docker Hub网站，注册账户，登录邮箱激活； (2) 登录Docker Hub
任务准备	本任务主要完成账户注册，需要保证手机号和电子邮箱的畅通，以接收验证码和注册验证。 创建账户时，要满足以下要求： (1) 手机号：一个常用或专用创建账户的手机号； (2) 电子邮箱：主流电子邮箱都可以，如QQ邮箱、网易邮箱等

📝 任务目标

知识目标	了解Linux、Docker Hub以及创建账号的过程
技能目标	能够熟练地对项目进行管理运用，并记录进展或问题
素质目标	耐心与严谨：在创建虚拟环境和项目的过程中，解决环境变量配置错误、依赖包版本错误等问题，提高个人的耐心与严谨的作风

🔧 任务实现

步骤1: 登录Docker Hub网站，注册账户，并登录电子邮箱激活，如图2-1-1～图2-1-4所示。

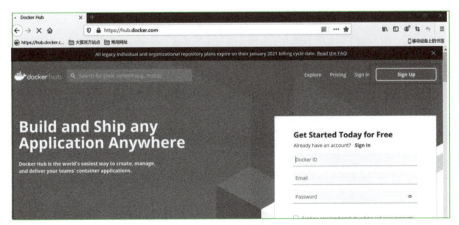

图 2-1-1　步骤图 1

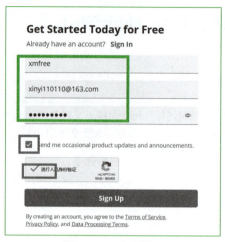

图 2-1-2　步骤图 2

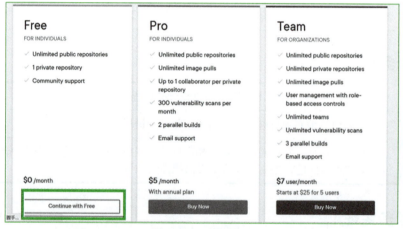

图 2-1-3　步骤图 3

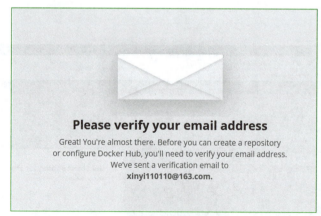

图 2-1-4　步骤图 4

项目 2 Docker 搭建与运维

步骤2： 登录Docker Hub，如图2-1-5所示。

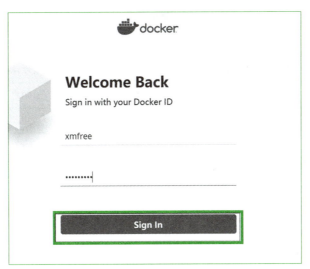

图 2-1-5 步骤图 6

```
[root@docker ~]# docker login docker.io
Login with your Docker ID to push and pull images from Docker Hub. If you
don't have a Docker ID, head over to https://hub.docker.com to create one.
  Username: xmfree  #输入注册的用户名
  Password:  #输入密码
WARNING! Your password will be stored unencrypted in /root/.docker/config.json.
Configure a credential helper to remove this warning. See
https://docs.docker.com/engine/reference/commandline/login/#credentials-store

Login Succeeded
```

学习笔记

考评记录

姓名			完成日期	
序号	考核内容		标准分	评分
1	成功注册账户，并登录邮箱激活		70	
2	登录账户		30	
	总评分		100	

任务实现心得：

任务实训

任务实训	注册仓库 Docker Hub 账户
任务目标	掌握如何创建 Docker Hub 账户并登录

单元 2　Docker 镜像管理 / 仓库管理

本单元包括两个任务：任务1是搜索镜像与下载镜像，包括搜索 alpine、Nginx 镜像，下载 alpine、Nginx 镜像，下载指定 tag；任务 2 是镜像的基本操作，包括查看本地镜像、给镜像打标签、推送镜像、删除镜像。

视频

Docker镜像管理、仓库管理

学习目标

通过本单元的学习，使学生掌握 Docker 镜像管理与仓库管理的基本知识，培养学生自主进行 Docker 镜像管理与仓库管理的能力。

任务 1　搜索镜像与下载镜像

任务描述

情境描述	容器除了运行其中应用外，基本不消耗额外的系统资源，使得应用的性能很高，同时系统的开销尽量小。传统虚拟机方式运行 10 个不同的应用就要启动10 个虚拟机，而 Docker 只需要启动 10 个隔离的应用即可。当运行容器时，如果使用的镜像在本地中不存在，Docker 就会自动从 Docker 镜像仓库中下载，默认是从 Docker Hub 公共镜像源下载。当程序员A使用Docker时，他发现没有alpine、Nginx 镜像，于是他准备进行下载
任务分解	分析上面的工作情境，将任务分解如下： (1) 搜索alpine、Nginx镜像； (2) 下载alpine镜像； (3) 下载Nginx镜像； (4) 下载指定tag
任务准备	在Docker使用之前要安装好相应的镜像。本任务主要进行的是alpine、Nginx镜像的搜索与下载，需要熟悉Linux的一些基本命令

任务目标

知识目标	掌握搜索以及下载镜像的知识
技能目标	学会使用Linux命令下载alpine、Nginx镜像
素质目标	耐心与严谨：在下载alpine、Nginx镜像的过程中，解决诸如环境变量配置错误、下载镜像包版本错误等问题，提高个人的耐心与严谨的作风

2-9

任务实现

步骤1： 搜索alpine、Nginx镜像，搜索结果如图2-2-1和图2-2-2所示。

```
[root@docker ~]# docker search alpine
NAME                                  DESCRIPTION                                     STARS   OFFICIAL   AUTOMATED
alpine                                A minimal Docker image based on Alpine Linux…   9532    [OK]
alpinelinux/docker-cli                Simple and lightweight Alpine Linux image wi…   6
alpinelinux/gitlab-runner             Alpine Linux gitlab-runner (supports more ar…   4
alpinelinux/alpine-gitlab-ci          Build Alpine Linux packages with Gitlab CI      3
alpinelinux/gitlab-runner-helper      Helper image container gitlab-runner-helper     2
grafana/alpine                        Alpine Linux with ca-certificates package in…   2
alpinelinux/gitlab                    Alpine Linux based Gitlab image                 2
alpinelinux/darkhttpd                                                                 1
alpinelinux/package-builder           Container to build packages for a repository    1
rancher/alpine-git                                                                    1
alpinelinux/golang                    Build container for golang based on Alpine L…   1
alpinelinux/docker-alpine                                                             0
alpinelinux/apkbuild-lint-tools       Tools for linting APKBUILD files in a CI env…   0
alpinelinux/build-base                Base image suitable for building packages wi…   0
alpinelinux/docker-abuild             Dockerised abuild                               0
alpinelinux/unbound                                                                   0
alpinelinux/mqtt-exec                                                                 0
alpinelinux/docker-compose            docker-compose image based on Alpine Linux      0
alpinelinux/git-mirror-syncd                                                          0
alpinelinux/mirror-status                                                             0
alpinelinux/ansible                   Ansible in docker                               0
alpinelinux/alpine-docker-gitlab      Gitlab running on Alpine Linux                  0
alpinelinux/alpine-drone-ci           Build Alpine Linux packages with drone CI       0
alpinelinux/aports-qa-bot             A Gitlab bot that gives feedback on aports m…   0
alpinelinux/rsyncd                                                                    0
```

图 2-2-1　搜索 alpine 镜像

```
[root@docker ~]# docker search nginx
NAME                                              DESCRIPTION                                     STARS   OFFICIAL   AUTOMATED
nginx                                             Official build of Nginx.                        17858   [OK]
linuxserver/nginx                                 An Nginx container, brought to you by LinuxS…   181
bitnami/nginx                                     Bitnami nginx Docker Image                      149                [OK]
ubuntu/nginx                                      Nginx, a high-performance reverse proxy & we…   71
bitnami/nginx-ingress-controller                  Bitnami Docker Image for NGINX Ingress Contr…   22                 [OK]
rancher/nginx-ingress-controller                                                                  11
kasmweb/nginx                                     An Nginx image based off nginx:alpine and in…   4
ibmcom/nginx-ingress-controller                   Docker Image for IBM Cloud Private-CE (Commu…   4
bitnami/nginx-ldap-auth-daemon                                                                    3
bitnami/nginx-exporter                                                                            3
circleci/nginx                                    This image is for internal use                  2
rancher/nginx                                                                                     2
rancher/nginx-ingress-controller-defaultbackend                                                   2
vmware/nginx                                                                                      2
rapidfort/nginx                                   RapidFort optimized, hardened image for NGINX   2
bitnami/nginx-intel                                                                               1
wallarm/nginx-ingress-controller                  Kubernetes Ingress Controller with Wallarm e…   1
vmware/nginx-photon                                                                               1
rancher/nginx-conf                                                                                0
ibmcom/nginx-ingress-controller-ppc64le           Docker Image for IBM Cloud Private-CE (Commu…   0
rapidfort/nginx-official                          RapidFort optimized, hardened image for NGIN…   0
ibmcom/nginx-ppc64le                              Docker image for nginx-ppc64le                  0
rancher/nginx-ssl                                                                                 0
rapidfort/nginx-ib                                RapidFort optimized, hardened image for NGIN…   0
continuumio/nginx-ingress-ws                                                                      0
```

图 2-2-2　搜索 Nginx 镜像

知识链接

镜像（Mirroring）是冗余的一种类型，一个磁盘上的数据在另一个磁盘上存在一个完全相同的副本即为镜像。镜像是一种文件存储形式，可以把许多文件做成一个镜像文件，与GHOST等程序放在一个盘里用GHOST等软件打开后，又恢复成许多文件。RAID 1和RAID 10使用的就是镜像。常见的镜像文件格式有ISO、BIN、IMG、TAO、DAO、CIF、FCD。

步骤2: 下载alpine镜像。

```
[root@docker ~]# docker pull alpine
Using default tag: latest
latest: Pulling from library/alpine
801bfaa63ef2: Pull complete
Digest: sha256:3c7497bf0c7af93428242d6176e8f7905f2201d8fc5861f45be7a346b5f23436
Status: Downloaded newer image for alpine:latest
       docker.io/library/alpine:latest
```

步骤3: 下载Nginx镜像。

```
[root@docker ~]# docker pull nginx
Using default tag: latest
latest: Pulling from library/nginx
a076a628af6f: Pull complete
0732ab25fa22: Pull complete
d7f36f6fe38f: Pull complete
f72584a26f32: Pull complete
7125e4df9063: Pull complete
Digest: sha256:10b8cc432d56da8b61b070f4c7d2543a9ed17c2b23010b43af434fd40e2ca4aa
Status: Downloaded newer image for nginx:latest
docker.io/library/nginx:latest
```

步骤4: 下载指定tag。

```
[root@docker ~]# docker pull alpine:3.12.3
3.12.3: Pulling from library/alpine
Digest: sha256:3c7497bf0c7af93428242d6176e8f7905f2201d8fc5861f45be7a346b5f23436
Status: Downloaded newer image for alpine:3.12.3
docker.io/library/alpine:3.12.3
```

学习笔记

考评记录

姓名			完成日期	
序号	考核内容		标准分	评分
1	搜索 alpine、Nginx 镜像		25	
2	下载 alpine 镜像		25	
3	下载 Nginx 镜像		25	
4	下载指定 tag		25	
	总评分		100	

任务实现心得：

任务实训	独立下载安装 alpine、Nginx 镜像
任务目标	能够独立下载所需要的镜像

任务 2　镜像的基本操作

选项	描述
ls	列出镜像
build	构建镜像来自 Dockerfile
history	查看镜像历史
inspect	显示一个或多个镜像详细信息
pull	发布容器端口到主机
push	推送一个镜像到镜像仓库
prune	移除未使用的镜像。没有被标记或被任何容器引用的
export	将文件系统作为一个 tar 归档文件导出到 STDOUT
import	从归档文件中创建镜像
tag	连接容器到一个网络
load	将镜像存储文件导入到本地镜像库
-v	绑定挂载一个卷
--restart	容器退出时重启策略，默认为 no

任务描述

情境描述	当程序员A完成Docker安装、注册仓库Docker Hub以及下载镜像后，他发现将进入镜像管理最重要的部分，那就是使用镜像，所以他准备了解镜像的基本操作
任务分解	分析以上工作情境，将任务分解如下： （1）查看本地镜像； （2）给镜像打标签； （3）推送镜像； （4）删除镜像
任务准备	（1）完成Docker安装； （2）注册仓库Docker Hub； （3）下载镜像； （4）掌握Linux基本操作

任务目标

知识目标	掌握镜像的基本操作
技能目标	能够独立利用镜像的一些操作，完成自己的任务
素质目标	耐心与严谨：通过对镜像基本操作的学习，提高个人的耐心与严谨的作风

步骤1: 查看本地镜像。

```
[root@docker ~]# docker images
REPOSITORY      TAG         IMAGE ID         CREATED         SIZE
alpine          3.12.3      389fef711851     2 weeks ago     5.58MB
alpine          latest      389fef711851     2 weeks ago     5.58MB
nginx           latest      ae2feff98a0c     3 weeks ago     133MB
```

步骤2: 给镜像打标签。

```
[root@docker ~]# docker tag 389fef711851 docker.io/xmfree/alpine:v3.12.3
[root@docker ~]# docker images
REPOSITORY        TAG         IMAGE ID         CREATED         SIZE
alpine            3.12.3      389fef711851     2 weeks ago     5.58MB
alpine            latest      389fef711851     2 weeks ago     5.58MB
xmfree/alpine     v3.12.3     389fef711851     2 weeks ago     5.58MB
nginx             latest      ae2feff98a0c     3 weeks ago     133MB
```

步骤3: 推送镜像。

```
[root@docker ~]# docker push docker.io/xmfree/alpine:v3.12.3
The push refers to repository [docker.io/xmfree/alpine]
777b2c648970: Mounted from library/alpine
v3.12.3: digest: sha256:074d3636ebda6dd446d0d00304c4454f468237fdacf08fb0eeac90bdbfa1bac7 size: 528
```

步骤4: 删除镜像。

```
[root@docker ~]# docker images
REPOSITORY        TAG         IMAGE ID         CREATED         SIZE
alpine            3.12.3      389fef711851     2 weeks ago     5.58MB
alpine            latest      389fef711851     2 weeks ago     5.58MB
xmfree/alpine     v3.12.3     389fef711851     2 weeks ago     5.58MB
nginx             latest      ae2feff98a0c     3 weeks ago     133MB
[root@docker ~]# docker rmi 389fef711851
Error response from daemon: conflict: unable to delete 389fef711851 (must be forced) - image is referenced in multiple repositories
[root@docker ~]# docker rmi -f 389fef711851
Untagged: alpine:3.12.3
Untagged: alpine@sha256:3c7497bf0c7af93428242d6176e8f7905f2201d8fc5861f45be7a346b5f23463
Untagged: xmfree/alpine:v3.12.3
Deleted: sha256:389fef7118515c70fd6c0e0d50bb75669942ea722ccb976507d7b087e54d5a23
Deleted: sha256:777b2c648970480f50f5b4d0af8f9a8ea798eea43dbcf40ce4a8c7118736bdcf
```

项目 2 Docker 搭建与运维

考评记录

姓名		完成日期	
序号	考核内容	标准分	评分
1	查看本地镜像	25	
2	给镜像打标签	25	
3	推送镜像	25	
4	删除镜像	25	
	总评分	100	

任务实现心得：

任务实训

任务实训	给镜像打标签
任务目标	掌握镜像的基本操作

单元 3　通过容器创建镜像与 Docker 容器管理

本单元包括两个任务：任务 1 是通过容器创建镜像，包括查看镜像、根据镜像启动容器、根据启动的容器创建新的镜像；任务 2 是 Docker 容器管理，包括运行容器、终止容器、启动容器、进入/退出容器、容器导出和导入、删除容器。

学习目标

通过本单元的学习，使学生掌握通过容器创建镜像与 Docker 容器管理的基本知识，培养学生自主进行通过容器创建镜像与 Docker 容器管理的能力。

任务 1　通过容器创建镜像

任务描述

视频
通过容器创建镜像

情境描述	程序员A发现在本地创建一个容器后，可以依据这个容器创建本地镜像，并可把这个镜像推送到Docker Hub中。于是他准备在自己的项目中也通过容器来创建镜像
任务分解	分析以上工作情境，将任务分解如下： (1) 查看镜像； (2) 根据镜像启动容器； (3) 根据启动的容器创建新的镜像
任务准备	(1) 掌握Linux基本操作； (2) 掌握在本地创建容器的方法； (3) 拥有一个完整的环境

任务目标

知识目标	掌握根据镜像启动容器的方法；掌握根据启动的容器创建新的镜像的方法
技能目标	能够独立通过容器创建镜像
素质目标	耐心及细心：通过练习使用容器创建镜像，培养学生面对问题以及解决问题的思维与耐心和细心

任务实现

步骤1： 查看镜像。

```
[root@docker ~]# docker images
```

```
REPOSITORY      TAG        IMAGE ID         CREATED        SIZE
nginx           latest     ae2feff98a0c     3 weeks ago    133MB
```

步骤2: 根据镜像启动容器。

```
[root@docker ~]# docker run -ti --name nginx -d nginx
eb49e0d3321a06ddc871251aaf1026dbcde56dc871d4508c9fe72dc0b079822d
[root@docker ~]# docker ps
CONTAINER ID      IMAGE              COMMAND                  CREATED
    STATUS      PORTS       NAMES
eb49e0d3321a     nginx            "/docker-entrypoint...."     4 seconds ago
Up 3 seconds   80/tcp      nginx
```

步骤3: 根据启动的容器创建新的镜像。

```
[root@docker ~]# docker commit -a "cool" -m "this is test" eb49e0d3321a nginx:1.0
sha256:caf7beb9af858d3c8884655ce468822fe7eeb643a85cc41fcb8f835d8974fcb7
[root@docker ~]# docker images
REPOSITORY      TAG        IMAGE ID         CREATED         SIZE
nginx           1.0        caf7beb9af85     3 seconds ago   133MB
nginx           latest     ae2feff98a0c     3 weeks ago     133MB
```

> **说明:**
>
> docker commit:从容器创建一个新的镜像。
>
> -a:提交的镜像作者;
>
> -c:使用 Dockerfile 指令创建镜像;
>
> -m:提交时的说明文字;
>
> -p:在 commit 时,将容器暂停。

学习笔记

考评记录

姓名			完成日期	
序号	考核内容		标准分	评分
1	查看镜像		30	
2	根据镜像启动容器		30	
3	根据启动的容器创建新的镜像		40	
	总评分		100	

任务实现心得：

任务实训

任务实训	修改根据镜像启动程序的参数
任务目标	掌握启动程序各个参数的含义

任务 2　Docker 容器管理

选项	描述
-i	交互式
-t	分配一个为终端
-d	运行容器到后台
-e	设置环境变量
-p	发布容器端口到主机
-P	发布容器所有 EXPOSE 的端口到宿主机随机端口
--name	指定容器名称
-h	指定容器主机名
--ip	指定容器 IP，只能用于自定义网络
--network	连接容器到一个网络
--mount	将文件系统附加到容器
-v	绑定挂载一个卷
--restart	容器退出时重启策略，默认为 no

视频

Docker容器管理

任务描述

情境描述	Docker可以很快速地创建一个生产环境。在复杂的生产环境中，运维开发者需要使用Docker的一些基本命令（如docker exec、docker log、docker attach、docker ps等）来维护生产环境，在安装好之后就需要掌握如何管理Docker。
任务分解	分析以上工作情境，将任务分解如下： （1）运行容器； （2）终止容器； （3）启动容器； （4）进入/退出容器； （5）容器导出和导入； （6）删除容器
任务准备	（1）掌握Linux基本操作； （2）记下Docker容器管理的操作指令； （3）拥有一个完整的环境

任务目标

知识目标	掌握各个参数与指令的含义
技能目标	学会独立使用各个指令进行Docker容器管理
素质目标	细心与严谨：通过让学生使用Docker容器管理的各个指令的练习，使学生拥有更高的细心与严谨性

任务实现

步骤1: 运行容器。

```
[root@docker ~]# docker run -dti --name nginx1.0 -p 80:80 nginx
9143ce0a31e7ae1b9f980aabab639529a148ea630a15df571e7076f6c5df9519
[root@docker ~]# docker ps
  CONTAINER ID        IMAGE              COMMAND                  CREATED
    STATUS            PORTS              NAMES
  9143ce0a31e7        nginx              "/docker-entrypoint...."   2 seconds ago
Up 2 seconds          0.0.0.0:80->80/tcp   nginx1.0
  eb49e0d3321a        nginx              "/docker-entrypoint...."   About an hour ago
Up About an hour      80/tcp             nginx
```

> **说明:**
>
> docker run:从容器创建一个新的镜像;
>
> -i:交互式操作;
>
> -t:终端;
>
> -d:容器在后台运行;
>
> -p:端口映射;
>
> --name:为容器指定一个名称。
>
> 如果指明-d 运行镜像,会返回容器id;如果不指明-d 运行镜像,会打印出 catalina.out 的日志,在按【Ctrl+C】组合键后,容器即停止运行。

步骤2: 终止容器。

```
[root@docker ~]# docker ps
  CONTAINER ID        IMAGE              COMMAND                       CREATED
STATUS                PORTS                   NAMES
  9143ce0a31e7        nginx              "/docker-entrypoint...."       7 minutes ago
Up 7 minutes          0.0.0.0:80->80/tcp      nginx1.0
  eb49e0d3321a        nginx              "/docker-entrypoint...."       About an hour ago
Up About an hour      80/tcp                  nginx
[root@docker ~]# docker stop 9143ce0a31e7 eb49e0d3321a
  9143ce0a31e7
  eb49e0d3321a
[root@docker ~]# docker ps
  CONTAINER ID        IMAGE          COMMAND         CREATED         STATUS PORTS        NAMES
[root@docker ~]# docker ps -a
  CONTAINER ID        IMAGE          COMMAND                                CREATED
STATUS                               PORTS                  NAMES
```

```
  9143ce0a31e7      nginx         "/docker-entrypoint...."      9 minutes ago
Exited (0) 4 seconds ago              nginx1.0
  eb49e0d3321a      nginx         "/docker-entrypoint...."     About an hour ago
Exited (0) 4 seconds ago              nginx
```

步骤3： 启动容器。

```
  [root@docker ~]# docker ps -a
  CONTAINER ID          IMAGE           COMMAND                CREATED
STATUS                          PORTS        NAMES
  9143ce0a31e7          nginx         "/docker-entrypoint...."      10 minutes ago
  Exited (0) About a minute ago              nginx1.0
  eb49e0d3321a          nginx         "/docker-entrypoint...."     About an hour ago
  Exited (0) About a minute ago              nginx
  [root@docker ~]# docker start 9143ce0a31e7
  9143ce0a31e7
  [root@docker ~]# docker ps
  CONTAINER ID          IMAGE           COMMAND                CREATED
STATUS              PORTS              NAMES
  9143ce0a31e7          nginx         "/docker-entrypoint...."      10 minutes ago
  Up 4 seconds    0.0.0.0:80->80/tcp    nginx1.0
```

> **说明：**
>
> docker container start [CONTAINER ID]：启动容器；
>
> docker container stop [CONTAINER ID]：终止容器；
>
> docker start $(docker ps -aq)：启动所有 Docker 容器。

步骤4： 进入/退出容器。

docker exec -it [CONTAINER ID]：bash进入容器。

代码如下：

```
  [root@docker ~]# docker ps
  CONTAINER ID          IMAGE           COMMAND                CREATED
STATUS          PORTS              NAMES
  9143ce0a31e7          nginx         "/docker-entrypoint...."      17 hours ago
Up17 hours    0.0.0.0:80->80/tcp    nginx1.0
  [root@docker ~]# docker exec -it 9143ce0a31e7 /bin/sh
  # ls
  bin   dev  docker-entrypoint.sh  home  lib64  mnt  proc  run  srv  tmp  var
```

```
boot  docker-entrypoint.d  etc  lib  media  opt root  sbin  sys  usr
# exit#退出容器
[root@docker ~]#
```

步骤5: 容器导出和导入。

(1) 容器导出。

```
#这样将导出容器快照到本地文件
docker export [CONTAINER ID] > [tar file]
```

代码如下：

```
[root@docker ~]# docker ps
CONTAINER ID      IMAGE        COMMAND                  CREATED
STATUS        PORTS                NAMES
9143ce0a31e7      nginx        "/docker-entrypoint...."   18 hours ago
Up17 hours   0.0.0.0:80->80/tcp    nginx1.0
[root@docker ~]# docker export 9143ce0a31e7 > nginx1.0.tar
[root@docker ~]# ls
anaconda-ks.cfg  nginx1.0.tar
```

(2) 容器导入。

```
#从容器快照文件中再导入为镜像
cat [tar file] | docker import - [name:tag]
```

代码如下：

```
[root@docker ~]# docker images
REPOSITORY    TAG        IMAGE ID         CREATED           SIZE
nginx         1.0        caf7beb9af85     19 hours ago      133MB
alpine        latest     389fef711851     3 weeks ago       5.58MB
nginx         latest     ae2feff98a0c     3 weeks ago       133MB
[root@docker ~]# cat nginx1.0.tar | docker import - nginx:1.0.1
sha256:6a652e89d56592369757470d6874e3567d4cda3b112f8cacba3e2b186486babd
[root@docker ~]# docker images
REPOSITORY    TAG        IMAGE ID         CREATED           SIZE
nginx         1.0.1      6a652e89d565     5 seconds ago     131MB
nginx         1.0        caf7beb9af85     19 hours ago      133MB
alpine        latest     389fef711851     3 weeks ago       5.58MB
nginx         latest     ae2feff98a0c     3 weeks ago       133MB
```

步骤6: 删除容器。

```
#删除终止状态的容器
docker rm [CONTAINER ID]
#删除所有处于终止状态的容器
docker container prune
#删除运行中的容器
docker rm -f [CONTAINER ID]
```

代码如下：

```
[root@docker ~]# docker ps
CONTAINER ID      IMAGE      COMMAND                  CREATED
STATUS         PORTS             NAMES
9143ce0a31e7     nginx     "/docker-entrypoint...."     18 hours ago
Up 18 hours    0.0.0.0:80->80/tcp    nginx1.0
[root@docker ~]# docker ps -a
CONTAINER ID      IMAGE      COMMAND                  CREATED
STATUS                PORTS                NAMES
9143ce0a31e7     nginx     "/docker-entrypoint...."     18 hours ago
Up 18 hours                 0.0.0.0:80->80/tcp    nginx1.0
eb49e0d3321a     nginx     "/docker-entrypoint...."     19 hours ago
Exited (0) 18 hours ago                       nginx
[root@docker ~]# docker rm eb49e0d3321a
eb49e0d3321a
[root@docker ~]# docker ps
CONTAINER ID      IMAGE      COMMAND                  CREATED
STATUS        PORTS              NAMES
9143ce0a31e7     nginx     "/docker-entrypoint...."     18 hours ago
Up 18 hours    0.0.0.0:80->80/tcp    nginx1.0
[root@docker ~]# docker rm -f 9143ce0a31e7
9143ce0a31e7
[root@docker ~]# docker ps
CONTAINER ID      IMAGE       COMMAND       CREATED        STATUS       PORTS    NAMES
[root@docker ~]#
```

学习笔记

任务考评

考评记录

姓名			完成日期	
序号	考核内容		标准分	评分
1	运行容器		10	
2	终止容器		10	
3	启动容器		20	
4	进入/退出容器		20	
5	容器导入和导出		20	
6	删除容器		20	
	总评分		100	

任务实现心得：

任务实训

任务实训	修改根据镜像启动程序的参数
任务目标	掌握启动程序各个参数的含义

项目② Docker 搭建与运维

单元 4　Docker 数据管理与 Docker 网络管理

本单元包括两个任务：任务 1 是 Docker 数据管理，包括创建两个容器、同时挂载数据卷容器、确认卷容器共享；任务 2 是 Docker 网络管理，包括创建一个 bridge 网络的容器、创建一个 host 网络的容器、创建一个 none 网络的容器、自定义网络模式。

学习目标

通过本单元的学习，使学生掌握 Docker 数据管理与 Docker 网络管理的基本知识，培养学生自主进行 Docker 数据管理与 Docker 网络管理的能力。

任务 1　Docker 数据管理之数据卷容器实践

任务描述

情境描述	如果需要在多个容器之间共享一些持续更新的数据，最简单的方式是使用数据卷容器。数据卷容器也是一个容器，它专门用来提供数据卷供其他容器挂载。程序员A在使用Docker的过程中，往往需要对数据进行持久化保存，或者需要在更多容器之间进行数据共享，他发现使用数据卷（Data Volumes）和数据卷容器（Data Volume Contain）就可以很好地解决这个问题。接下来他进行了数据卷容器实践
任务分解	分析以上工作情境，将任务分解如下： （1）创建两个容器，同时挂载数据卷容器； （2）确认卷容器共享
任务准备	（1）掌握Linux基本操作； （2）了解数据卷以及数据卷容器； （3）拥有一个完整的环境

视频

Docker数据管理

任务目标

知识目标	掌握创建两个容器，并且同时挂载数据卷容器的方法
技能目标	能够独立完成数据卷实践操作
素质目标	耐心与严谨：在进行数据卷容器实践的过程中，提高个人的耐心与严谨的作风

任务实现

步骤1： 创建两个容器，同时挂载数据卷容器。

docker run --volumes-from [数据卷容器id/name] -dti --name[容器名字][镜像名称]　[命令(可选)]

2-25

创建v01容器：

```
[root@docker ~]# docker run --volumes-from v-data1 -dti --name v01 nginx
1390a43c879a9e715cffdb76a96eafd188e7e3ef1bff0ef0d2a5e9ec6d089375
```

创建v02容器：

```
[root@docker ~]# docker run --volumes-from v-data1 -dti --name v02 nginx
7685c28d94753441721be5741064685092d6cda7d46beb44ef46080ae678545e

[root@docker ~]# docker ps -a
 CONTAINER ID        IMAGE         COMMAND                  CREATED             STATUS
PORTS         NAMES
  7685c28d9475      nginx        "/docker-entrypoint.…"    4 hours ago      Up 4 hours
 80/tcp       v02
  1390a43c879a      nginx        "/docker-entrypoint.…"    4 hours ago      Up 4 hours
 80/tcp       v01
  6da6aff079ab      nginx        "/docker-entrypoint.…"    4 hours ago      Created
v-data1
```

知识链接

数据卷就是将宿主机的某个目录映射到容器中，作为数据存储的目录，然后就可以在宿主机对数据进行存储。

（1）数据卷可以在容器之间共享和重用，本地与容器间传递数据更高效；

（2）对数据卷的修改会立即生效，容器内部与本地目录均可；

（3）对数据卷的更新不会影响镜像，对数据与应用进行了解耦操作；

（4）数据卷默认会一直存在，即使容器被删除。

步骤2： 确认卷容器共享。

进入v01容器，操作数据卷容器：

```
[root@docker ~]# docker exec -it 1390a43c879a /bin/bash
root@1390a43c879a:/# ls /data1
root@1390a43c879a:/# echo 'v01' >/data1/v01.txt
root@1390a43c879a:/# exit
exit
```

进入v02容器，确认数据卷：

```
[root@docker ~]# docker exec -it 7685c28d9475 /bin/bash
root@7685c28d9475:/# echo 'v02'>/data1/v02.txt
root@7685c28d9475:/# ls /data1/
v01.txt  v02.txt
root@7685c28d9475:/# exit
exit
```

考评记录

姓名			完成日期	
序号	考核内容		标准分	评分
1	创建两个容器，同时挂载数据卷容器		50	
2	确认卷容器共享		50	
	总评分		100	

任务实现心得：

任务实训

任务实训	创建五个容器，并同时挂载数据卷容器
任务目标	掌握数据卷容器的使用

项目 2 Docker 搭建与运维

2-27

任务 2　Docker 网络管理

视频
Docker网络管理

任务描述

情境描述	程序员A在使用docker run命令创建Docker容器时，可以用--net选项指定容器的网络模式。Docker有以下四种网络模式： （1）host模式，使用--net=host指定； （2）container模式，使用--net=container:NAME_or_ID指定； （3）none模式，使用--net=none指定； （4）bridge模式，使用--net=bridge指定，默认设置。 程序员A想了解一下Docker的各个网络模式
任务分解	分析以上工作情境，将任务分解如下： （1）创建一个bridge网络的容器； （2）创建一个host网络的容器； （3）创建一个none网络的容器； （4）自定义网络模式
任务准备	（1）掌握Linux基本操作； （2）拥有一个完整的环境

任务目标

知识目标	了解Docker网络管理的四种模式
技能目标	能够独立创建四种不同网络模式的容器
素质目标	细心与耐心：在创建四种不同模式的过程中，提高学生的细心与耐心

任务实现

步骤1： 创建一个bridge网络的容器。

```
[root@docker ~]# docker run -it busybox
Unable to find image 'busybox:latest' locally
latest: Pulling from library/busybox
4c892f00285e: Pull complete
Digest: sha256:e1488cb900233d035575f0a7787448cb1fa93bed0ccc0d4efc1963d7d72a8f17
Status: Downloaded newer image for busybox:latest
/ #
```

知识链接

当Docker进程启动时，会在主机上创建一个名为docker0的虚拟网桥，此主机上启动的Docker容器会连接到这个虚拟网桥上。虚拟网桥的工作方式和物理交换机类似，这样主机上的所有容器就通过交换机连在了一个二层网络中。

从docker0子网中分配一个IP给容器使用，并设置docker0的IP地址为容器的默认网关。在主机上创建一对虚拟网卡veth pair设备，Docker将veth pair设备的一端放在新创建的容器中，并命名为eth0（容器的网卡）；另一端放在主机中，以veth×××这样的名字命名，并将这个网络设备加入docker0网桥中。可以通过brctl show命令查看。

bridge模式是Docker的默认网络模式，不写--net参数，就是bridge模式。使用docker run -p时，Docker实际是在iptables做了DNAT规则，实现端口转发功能。可以使用iptables -t nat -vnL查看。

步骤2: 创建一个host网络的容器。

```
[root@docker ~]# docker run -it --network host busybox
/ # ifconfig
docker0   Link encap:Ethernet  HWaddr 02:42:77:58:6B:70
          inet addr:172.17.0.1  Bcast:172.17.255.255  Mask:255.255.0.0
          inet6 addr: fe80::42:77ff:fe58:6b70/64 Scope:Link
          UP BROADCAST MULTICAST  MTU:1500  Metric:1
          RX packets:0 errors:0 dropped:0 overruns:0 frame:0
          TX packets:5 errors:0 dropped:0 overruns:0 carrier:0
          collisions:0 txqueuelen:0
          RX bytes:0 (0.0 B)  TX bytes:438 (438.0 B)

ens33     Link encap:Ethernet  HWaddr 00:0C:29:4E:AF:EE
          inet addr:192.168.137.3  Bcast:192.168.137.255  Mask:255.255.255.0
          inet6 addr: fe80::20c:29ff:fe4e:afee/64 Scope:Link
          UP BROADCAST RUNNING MULTICAST  MTU:1500  Metric:1
          RX packets:839973 errors:0 dropped:0 overruns:0 frame:0
          TX packets:261509 errors:0 dropped:0 overruns:0 carrier:0
          collisions:0 txqueuelen:1000
          RX bytes:1073840334 (1.0 GiB)  TX bytes:29048386 (27.7 MiB)

lo        Link encap:Local Loopback
          inet addr:127.0.0.1  Mask:255.0.0.0
          inet6 addr: ::1/128 Scope:Host
          UP LOOPBACK RUNNING  MTU:65536  Metric:1
          RX packets:5947 errors:0 dropped:0 overruns:0 frame:0
          TX packets:5947 errors:0 dropped:0 overruns:0 carrier:0
          collisions:0 txqueuelen:1
          RX bytes:2251874 (2.1 MiB)  TX bytes:2251874 (2.1 MiB)
```

通过以上实验得知，该容器的网络与宿主机一样，直接使用Dockerhost的网络，最大的好处就是性能。如果容器对网络传输效率有较高要求，则可以选择host网络。当然，不便之处就是会牺牲一些灵活性，比如要考虑端口冲突问题，Dockerhost上已经使用的端口就不能再用了。

如果使用host模式启动容器，那么该容器将不会获得独立的Network Namespace，而是和宿主机共用一个Network Namespace。容器将不会虚拟出自己的网卡、配置自己的IP等，而是使用宿主机的IP和端口。但是，容器的其他方面，如文件系统、进程列表等还是和宿主机隔离的。

使用host模式的容器可以直接使用宿主机的IP地址与外界通信，容器内部的服务端口也可以使用宿主机的端口，不需要进行NAT。host最大的优点是网络性能比较好；缺点是网络的隔离性不好。

步骤3： 创建一个none网络的容器。

```
[root@docker ~]# docker run -it --network none busybox
/ # ifconfig
lo        Link encap:Local Loopback
          inet addr:127.0.0.1  Mask:255.0.0.0
          UP LOOPBACK RUNNING  MTU:65536  Metric:1
          RX packets:0 errors:0 dropped:0 overruns:0 frame:0
          TX packets:0 errors:0 dropped:0 overruns:0 carrier:0
          collisions:0 txqueuelen:1
          RX bytes:0 (0.0 B)  TX bytes:0 (0.0 B)

/ #
```

例如，某个容器的唯一用途是生产随机密码，就可以放到none网络中避免密码被窃取。使用none模式，Docker容器拥有自己的Network Namespace，但是，并不对Docker容器进行任何网络配置。也就是说，这个Docker容器没有网卡、IP、路由等信息。需要程序员自行为Docker容器添加网卡、配置IP等。

这种网络模式下容器只有lo回环网络，没有其他网卡。none模式可以在容器创建时通过--network=none来指定。这种类型的网络无法联网，封闭的网络能很好地保证容器的安全性。

步骤4： 自定义网络模式。

创建my_net网络，系统默认IP地址段自动向后递增。

```
[root@docker ~]# docker network create --driver bridge my_net
4b2f6fc053a2e255d8f27c1a9d81cb44406a2c5ed1aeed8f7c95f806c5cac5e4
[root@docker ~]# ifconfig
br-4b2f6fc053a2: flags=4099<UP,BROADCAST,MULTICAST>  mtu 1500
        inet 172.18.0.1  netmask 255.255.0.0  broadcast 172.18.255.255
        ether 02:42:02:8e:37:d1  txqueuelen 0  (Ethernet)
        RX packets 0  bytes 0 (0.0 B)
        RX errors 0  dropped 0  overruns 0  frame 0
        TX packets 0  bytes 0 (0.0 B)
        TX errors 0  dropped 0 overruns 0  carrier 0  collisions 0
```

```
docker0: flags=4099<UP,BROADCAST,MULTICAST>  mtu 1500
        inet 172.17.0.1  netmask 255.255.0.0  broadcast 172.17.255.255
        inet6 fe80::42:77ff:fe58:6b70  prefixlen 64  scopeid 0x20<link>
        ether 02:42:77:58:6b:70  txqueuelen 0  (Ethernet)
        RX packets 0  bytes 0 (0.0 B)
        RX errors 0  dropped 0  overruns 0  frame 0
        TX packets 5  bytes 438 (438.0 B)
        TX errors 0  dropped 0 overruns 0  carrier 0  collisions 0

ens33: flags=4163<UP,BROADCAST,RUNNING,MULTICAST>  mtu 1500
        inet 192.168.137.3  netmask 255.255.255.0  broadcast 192.168.137.255
        inet6 fe80::20c:29ff:fe4e:afee  prefixlen 64  scopeid 0x20<link>
        ether 00:0c:29:4e:af:ee  txqueuelen 1000  (Ethernet)
        RX packets 840094  bytes 1073852046 (1.0 GiB)
        RX errors 0  dropped 0  overruns 0  frame 0
        TX packets 261572  bytes 29060974 (27.7 MiB)
        TX errors 0  dropped 0 overruns 0  carrier 0  collisions 0

lo: flags=73<UP,LOOPBACK,RUNNING>  mtu 65536
        inet 127.0.0.1  netmask 255.0.0.0
        inet6 ::1  prefixlen 128  scopeid 0x10<host>
        loop  txqueuelen 1  (Local Loopback)
        RX packets 5947  bytes 2251874 (2.1 MiB)
        RX errors 0  dropped 0  overruns 0  frame 0
        TX packets 5947  bytes 2251874 (2.1 MiB)
        TX errors 0  dropped 0 overruns 0  carrier 0  collisions 0
```

使用自定义网络创建容器。

```
[root@docker ~]# docker run --network my_net -it busybox
/ # ifconfig
eth0      Link encap:Ethernet   HWaddr 02:42:AC:12:00:02
          inet addr:172.18.0.2  Bcast:172.18.255.255  Mask:255.255.0.0
          UP BROADCAST RUNNING MULTICAST  MTU:1500  Metric:1
          RX packets:11 errors:0 dropped:0 overruns:0 frame:0
          TX packets:0 errors:0 dropped:0 overruns:0 carrier:0
          collisions:0 txqueuelen:0
          RX bytes:946 (946.0 B)  TX bytes:0 (0.0 B)

lo        Link encap:Local Loopback
          inet addr:127.0.0.1  Mask:255.0.0.0
          UP LOOPBACK RUNNING  MTU:65536  Metric:1
          RX packets:0 errors:0 dropped:0 overruns:0 frame:0
          TX packets:0 errors:0 dropped:0 overruns:0 carrier:0
          collisions:0 txqueuelen:1
          RX bytes:0 (0.0 B)  TX bytes:0 (0.0 B)
```

任务考评

考评记录

姓名			完成日期	
序号	考核内容		标准分	评分
1	创建一个 bridge 网络的容器		25	
2	创建一个 host 网络的容器		25	
3	创建一个 none 网络的容器		25	
4	自定义网络模式		25	
	总评分		100	

任务实现心得：

任务实训

任务实训	分别独立创建四种不同网络模式的容器
任务目标	掌握 Docker 网络管理的四种不同模式

项目 2　Docker 搭建与运维

单元 5　Dockerfile 使用

本单元包括两个任务：任务 1 是自定义镜像，包括定制 centos、执行构建、验证是否支持所需功能、查看镜像变更历史；任务 2 是自定义 Tomcat9 镜像，包括环境准备、编辑 mydockerfile 文件、在 root 目录下执行构建、运行查看效果。

视　频

Dockerfile 使用

学习目标

通过本单元的学习，使学生掌握 Dockerfile 使用的基本知识，培养学生自主进行自定义镜像与自定义 Tomcat9 镜像的能力。

任务 1　自定义镜像

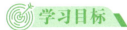

任务描述

情境描述	Dockerfile 是用来构建 Docker 镜像的构建文件，是由一系列命令和参数构成的脚本。每条保留字指令都必须为大写字母，且后面要跟随至少一个参数，指令从上到下顺序执行，"#"表示注释。每条指令都会创建一个新的镜像层，并对镜像进行提交。Dockerfile 可以很方便地基于已有镜像创建新的镜像。 Docker 执行 Dockerfile 的大致流程如下： （1）Docker 从基础镜像运行一个容器； （2）执行一条指令并对容器作出修改； （3）执行类似Docker commit 的操作提交一个新的镜像层； （4）Docker 再基于刚提交的镜像运行一个新容器； （5）执行 Dockerfile 中的下一条指令，直到所有指令都执行完成。 Dockerfile文件中包含若干命令，每个命令都会创建一个新的层。Dockerfile创建的层数不可以超过127层。Docker Hub 中 99% 的镜像都是通过在 base 镜像中安装和配置需要的软件构建出来的，如FROM scratch、FROM centos。程序员B在程序员A之前搭好环境的基础上进行了自定义镜像
任务分解	分析以上工作情境，将任务分解如下： （1）定制 centos； （2）执行构建； （3）验证是否支持所需功能； （4）查看镜像变更历史
任务准备	（1）掌握Linux基本操作； （2）拥有一个完整的环境； （3）登录后的默认路径、vim 编辑器、查看网络配置 ifconfig 支持

2-33

任务目标

知识目标	（1）掌握Dockerfile指令； （2）学会如何自定义镜像
技能目标	能够独立创建自定义镜像
素质目标	耐心与细心：在自定义镜像过程中，针对可能出现的问题，要有足够的耐心与细心

任务实现

步骤1： 自定义CentOS镜像。

使自己的镜像具备：登录后的默认路径、vim 编辑器、查看网络配置 ifconfig支持。

```
[root@docker ~]# cat /root/Dockerfile
FROM centos
MAINTAINER cool
ENV MYPATH /usr/local
WORKDIR $MYPATH
RUN yum -y install vim&&yum -y install net-tools
EXPOSE 80
CMD echo $MYPATH
CMD echo "success--------------ok"
CMD /bin/bash
```

知识链接

Dockerfile指令见下表。

指 令	描 述
FROM	构建新镜像是基于哪个镜像
MAINTAINER	镜像维护者姓名或邮箱地址
RUN	构建镜像时运行的 shell 命令
COPY	复制文件或目录到镜像中
ENV	设置环境变量
USER	为 RUN、CMD、ENTRYPOINT 执行命令指定运行用户
EXPOSE	声明容器运行的服务端口
HEALTHCHECK	容器中服务健康检查
WORKDIR	为 RUN、CMD、ENTRYPOINT、COPY 和 ADD 设置工作目录
ENTRYPOINT	运行容器时，如果有多个 ENTRYPOINT 指令，最后一个生效
CMD	运行容器时执行，如果有多个 CMD 指令，最后一个生效

步骤2： 执行构建。

```
[root@docker ~]# docker build -t mycentos:v1 .
Sending build context to Docker daemon   25.6kB
Step 1/9 : FROM centos
latest: Pulling from library/centos
7a0437f04f83: Pull complete
Digest: sha256:5528e8b1b1719d34604c87e11dcd1c0a20bedf46e83b5632cdeac91b8c04efc1
Status: Downloaded newer image for centos:latest
 ---> 300e315adb2f
Step 2/9 : MAINTAINER cool
 ---> Running in 1f5732a443eb
Removing intermediate container 1f5732a443eb
 ---> ff35396e9568
Step 3/9 : ENV MYPATH /usr/local
 ---> Running in b5cd8568a706
Removing intermediate container b5cd8568a706
 ---> 74b8e3fce28c
Step 4/9 : WORKDIR $MYPATH
 ---> Running in 962685eaae5a
Removing intermediate container 962685eaae5a
 ---> 7627ce2219b5
Step 5/9 : RUN yum -y install vim && yum -y install net-tools
Successfully built 93d37fc08d97
Successfully tagged mycentos:v1
```

步骤3： 验证是否支持所需功能。

```
[root@docker ~]# docker run -it mycentos:v1
[root@a007c1e857a7 local]# pwd
/usr/local
[root@a007c1e857a7 local]# vim --version
VIM - Vi IMproved 8.0 (2016 Sep 12, compiled Jun 18 2020 15:49:08)
Included patches: 1-1763
```

步骤4： 查看镜像变更历史。

```
[root@docker ~]# docker history mycentos:v1
IMAGE           CREATED         CREATED BY                                       SIZE    COMMENT
93d37fc08d97    30 minutes ago  /bin/sh -c #(nop)  CMD ["/bin/sh" "-c" "/bin...  0B
a45f59237d50    30 minutes ago  /bin/sh -c #(nop)  CMD ["/bin/sh" "-c" "echo...  0B
aca67303d7b9    30 minutes ago  /bin/sh -c #(nop)  CMD ["/bin/sh" "-c" "echo...  0B
```

2b100c7a944e	30 minutes ago	/bin/sh -c #(nop)	EXPOSE 80	0B
5d3a69c5fd7d	30 minutes ago	/bin/sh -c yum -y	install vim && yum -y inst...	58.8MB
7627ce2219b5	30 minutes ago	/bin/sh -c #(nop)	WORKDIR /usr/local	0B
74b8e3fce28c	30 minutes ago	/bin/sh -c #(nop)	ENV MYPATH=/usr/local	0B
ff35396e9568	30 minutes ago	/bin/sh -c #(nop)	MAINTAINER cool	0B
300e315adb2f	8 weeks ago	/bin/sh -c #(nop)	CMD ["/bin/bash"]	0B
\<missing\>	8 weeks ago	/bin/sh -c #(nop)	LABEL org.label-schema.sc...	0B
\<missing\>	8 weeks ago	/bin/sh -c #(nop)	ADD file:bd7a2aed6ede423b7...	209MB

学习笔记

任务考评

考评记录

姓名		完成日期	
序号	考核内容	标准分	评分
1	定制 centos	25	
2	执行构建	25	
3	验证是否支持所需功能	25	
4	查看镜像变更历史	25	
	总评分	100	

任务实现心得：

任务实训

任务实训	修改根据镜像启动程序的参数
任务目标	掌握启动程序的各个参数的含义

任务 2　自定义 Tomcat 9 镜像

任务描述

情境描述	程序员A发现Dockerfile可以很方便地基于已有镜像创建新的镜像，并在学习如何自定义镜像之后，准备进行自定义Tomcat9镜像
任务分解	分析上面的工作情境，将任务分解如下： （1）环境准备； （2）编辑mydockerfile文件； （3）在root目录下执行构建； （4）运行查看效果
任务准备	（1）掌握Linux基本操作； （2）拥有一个完整的环境； （3）下载 Tomcat 9.0.41及jdk

任务目标

知识目标	掌握如何自定义Tomcat 9镜像
技能目标	（1）能够独立编辑mydockerfile文件； （2）能够独立在root目录下执行构建； （3）能够独立运行查看效果
素质目标	耐心与细心：在自定义Tomcat9镜像过程中，对于代码报错问题能够耐心地查找和解决，对于繁多的代码能够细心地查看代码逻辑

任务实现

步骤1： 环境准备。

通过网络搜索下载Tomcat 9.0.41及jdk，下载到本地后，进行下列操作：

```
[root@docker ~]# cd /root
[root@docker ~]# tar xf apache-tomcat-9.0.41.tar.gz
[root@docker ~]# tar xf jdk-8u281-linux-x64.tar.gz
```

```
[root@docker ~]# ls
anaconda-ks.cfg         apache-tomcat-9.0.41.tar.gz    jdk1.8.0_281
apache-tomcat-9.0.41    Dockerfile                     jdk-8u281-linux-x64.tar.gz
```

步骤2: 编辑mydockerfile文件。

```
[root@docker ~]# cat mydockerfile
FROM centos
MAINTAINER cool
# 把宿主机当前目录下的jdk1.8.0_281复制到容器/usr/local/路径下
COPY jdk1.8.0_281/ /usr/local/jdk1.8.0_281/
# 把宿主机当前目录下的Tomcat添加到容器/usr/local/路径下
ADD apache-tomcat-9.0.41.tar.gz /usr/local/
# 安装vim编辑器
RUN yum -y install vim
# 设置工作访问时的WORKDIR路径，登录落脚点
ENV MYPATH /usr/local
WORKDIR $MYPATH
# 配置Java与Tomcat 环境变量
ENV JAVA_HOME /usr/local/jdk1.8.0_281
ENV CLASSPATH $JAVA_HOME/lib/dt.jar:$JAVA_HOME/lib/tools.jar
ENV CATALINA_HOME /usr/local/apache-tomcat-9.0.41
ENV CATALINA_BASE /usr/local/apache-tomcat-9.0.41
ENV PATH $PATH:$JAVA_HOME/bin:$CATALINA_HOME/lib:$CATALINA_HOME/bin
# 容器运行时监听的端口
EXPOSE  8080
# 启动时运行Tomcat
# ENTRYPOINT ["/usr/local/apache-tomcat-9.0.41/bin/startup.sh" ]
CMD ["/usr/local/apache-tomcat-9.0.41/bin/catalina.sh","run"]
```

步骤3: 在root目录下执行构建。

```
[root@docker ~]# docker build -f mydockerfile -t tomcat:v1 .
```

步骤4: 运行查看效果，如图2-5-1所示。

```
[root@docker ~]# docker run -d -p 8080:8080 --privileged=true --name tomcat9 -v /tmp/tomcat9logs/:/usr/local/apache-tomcat-9.0.41/logs tomcat:v1
4337a0e76e5dedd8123918dfa7e95bf93d69a2b6dc428044c553a2774c98999e
浏览器访问ip+映射的端口
http://192.168.137.3:8080/
```

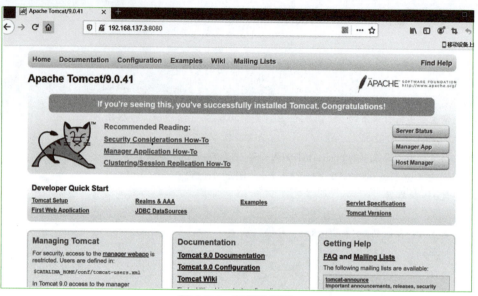

图 2-5-1　运行效果图

学习笔记

项目 2　Docker 搭建与运维

任务考评

考评记录

姓名		完成日期	
序号	考核内容	标准分	评分
1	环境准备	25	
2	编辑 mydockerfile 文件	25	
3	在 root 目录下执行构建	25	
4	运行查看效果	25	
	总评分	100	

任务实现心得：

任务实训

任务实训	掌握如何自定义 Tomcat9 镜像
任务目标	可以独立编辑 mydockerfile 文件

2-41

项目 3
ELK 日志分析系统

本项目主要介绍了 ELK 日志分析系统。Elasticsearch 是一个基于 Lucene 的搜索服务器。它提供了一个分布式多用户能力的全文搜索引擎，基于 RESTful Web 接口。Elasticsearch 是用 Java 语言开发的，并作为 Apache 许可条款下的开放源码发布，是一种流行的企业级搜索引擎。Elasticsearch 用于云计算中，能够达到实时搜索、稳定、可靠、快速、安装使用方便。本项目共分为三个单元，分别介绍了 Elasticsearch 安装与配置、监测集群状态和部署 Kibana、Logstash 的部署和收集日志。

ELK介绍

单元 1　Elasticsearch 安装与配置

本单元包括三个任务：任务 1 是基础环境配置，包括环境准备、修改主机名、配置 host 文件、安装 JDK；任务 2 是 Elasticsearch 安装，包括远程复制文件与安装方法介绍；任务 3 是配置 Elasticsearch，包括配置 Elasticsearch 的配置文件与启动服务。

学习目标

通过本单元的学习，使学生掌握基础环境配置和 Elasticsearch 安装与配置的基本知识，培养学生自主进行基础环境配置和 Elasticsearch 安装与配置的能力。

任务 1　基础环境配置

任务描述

基础环境配置

情境描述	如果需要查看某些服务为什么错误，可以直接使用grep等命令；如果查看规模较大，日志较多，此方法效率就低了很多。现在对待大规模的日志，解决思路是建立集中式日志收集系统，将所有节点上的日志统一收集、管理、访问。 一个完整的集中式日志系统需要包含以下几个主要特点： （1）收集：能够采集多种来源的日志数据； （2）传输：能够稳定的把日志数据传输到中央系统； （3）存储：能够存储日志数据； （4）分析：能够支持 UI 分析； （5）警告：能够提供错误报告和监控机制。 ELK提供一整套的解决方案，并且都是开源软件，它们之间相互配合，完美衔接，高效地满足了很多场合的应用。程序员A想下载Kibana工具，需要先了解基础环境配置
任务分解	分析上面的工作情境，将任务分解如下： （1）环境准备； （2）修改主机名； （3）配置host文件； （4）安装JDK
任务准备	掌握Linux系统基本操作

任务目标

知识目标	了解Kibana所需的基础环境
技能目标	（1）学会如何修改主机名； （2）学会如何配置host文件； （3）学会JDK的安装
素质目标	耐心与细心：在host文件配置过程中，针对代码报错问题能够耐心地查找和解决问题，对于繁多的代码能够细心地查看代码逻辑

任务实现

步骤1： 环境准备。

规划三个节点，其中一个作为主节点，两个作为数据节点。

节点IP	节点规划	主机名
192.168.40.11	Elasticsearch+Kibana（主）	elk-1
192.168.40.12	Elasticsearch+Logstash（数据）	elk-2
192.168.40.13	Elasticsearch（数据）	elk-3

知识链接

ELK是三个开源软件的缩写，分别为Elasticsearch、Logstash和Kibana。Elasticsearch是个开源分布式搜索引擎，提供搜集、分析、存储数据三大功能。它的特点有分布式、零配置、自动发现、索引自动分片、索引副本机制、restful风格接口、多数据源、自动搜索负载等。现在还新增了一个Beats，它是一个轻量级的日志收集处理工具（Agent）。Beats占用资源少，适合于在各个服务器上搜集日志后传输给Logstash，官方也推荐此工具。目前，由于原本的ELK Stack成员中加入了Beats工具，因此已改名为Elastic Stack。Logstash主要是用来日志的搜集、分析、过滤日志的工具，支持大量的数据获取方式。一般工作方式为C/S架构，Client端安装在需要收集日志的主机上，Server端负责将收到的各节点日志进行过滤、修改等操作，再一并发往Elasticsearch上去。Kibana可以为Logstash和ElasticSearch提供日志分析Web界面，可以帮助汇总、分析和搜索重要数据日志。Beats是一个轻量级日志采集器，Beats家族有六个成员，早期的ELK架构中使用Logstash收集、解析日志，但是Logstash对内存、CPU、IO等资源消耗比较高。相比Logstash，Beats所占系统的CPU和内存几乎可以忽略不计。

步骤2： 修改主机名。

使用hostnamectl命令修改三个主机名，以做区分。
elk-1节点：

```
[root@localhost ~]# hostnamectl set-hostname elk-1
// 修改完后按【Ctrl+D】组合键退出后重新连接
[root@elk-1 ~]#
```

elk-2节点：

```
[root@localhost ~]# hostnamectl set-hostname elk-2
[root@elk-2 ~]#
```

elk-3节点：

```
[root@localhost ~]# hostnamectl set-hostname elk-3
[root@elk-3 ~]#
```

步骤3： 配置hosts文件。

三个节点配置相同（以elk-1节点为例），命令如下：
elk-1节点：

```
[root@elk-1 ~]# vi /etc/hosts
[root@elk-1 ~]# cat /etc/hosts
127.0.0.1    localhostlocalhost.localdomain localhost4 localhost4.localdomain4
::1          localhostlocalhost.localdomain localhost6 localhost6.localdomain6
192.168.40.11 elk-1
192.168.40.12 elk-2
192.168.40.13 elk-3
```

步骤4： 安装JDK。

部署ELK环境需要JDK 1.8以上的JDK版本软件环境，这里使用openjdk1.8，三节点全部安装（以elk-1节点为例），命令如下：
elk-1节点：

```
[root@elk-1 ~]# yum install -y java-1.8.0-openjdk java-1.8.0-openjdk-devel
...
[root@elk-1 ~]# java -version
openjdk version "1.8.0_242"
OpenJDK Runtime Environment (build 1.8.0_242-b08)
OpenJDK 64-Bit Server VM (build 25.242-b08, mixed mode)
```

知识链接

JDK是Java语言的软件开发工具包，主要用于移动设备、嵌入式设备上的Java应用程序。JDK是整个Java开发的核心，它包含了Java的运行环境（JVM+Java系统类库）和Java工具。

学习笔记

考评记录

姓名		完成日期	
序号	考核内容	标准分	评分
1	环境准备	25	
2	修改主机名	25	
3	配置 hosts 文件	25	
4	安装 JDK	25	
	总评分	100	

任务实现心得：

任务实训

任务实训	通过修改 host 中的 IP 地址，看看会发生什么
任务目标	掌握 host 配置文件中的 IP 地址的含义

任务 2 Elasticsearch 安装

● 视频

Elasticsearch 安装

任务描述

情境描述	程序员B查找了相关资料，了解到Elasticsearch是个开源分布式搜索引擎，提供搜集、分析、存储数据三大功能。它的特点有分布式、零配置、自动发现、索引自动分片、索引副本机制、restful风格接口、多数据源、自动搜索负载等。程序员A想进行分布式多用户能力的全文搜索，但是试了很多软件都效果不佳，项目经理提议使用Elasticsearch。它提供了一个分布式多用户能力的全文搜索引擎，基于RESTful Web接口。Elasticsearch是用Java语言开发的，并作为Apache许可条款下的开放源码发布，是一种流行的企业级搜索引擎，很适合当下公司正在开发的任务，所以他决定动手安装这个搜索引擎
任务分解	分析上面的工作情境，将任务分解如下： (1) 远程复制文件； (2) 安装方法介绍
任务准备	掌握Linux系统基本操作，以及节点查看方式

任务目标

知识目标	了解Elasticsearch的安装过程
技能目标	掌握Elasticsearch的安装和节点的查看
素质目标	耐心与细心：在Elasticsearch的安装过程中，针对安装过程出现的问题耐心地查找和解决问题

任务实现

步骤1： 远程复制文件。

```
[root@elk-1 ~]# scp elasticsearch-6.0.0.rpm elk-3:/root/
The authenticity of host 'elk-3 (192.168.40.13)' can't be established.
ECDSA key fingerprint is f3:72:41:05:79:cd:52:9b:a6:98:f0:5b:e8:5f:26:3d.
Are you sure you want to continue connecting (yes/no)? y
//第一次连接会询问是否确定连接，第二次连接就会只让你输入密码
Please type 'yes' or 'no': yes
Warning: Permanently added 'elk-3,192.168.40.13' (ECDSA) to the list of known hosts.
root@elk-3's password:
// 连接的机器的密码，就是elk-3这台机器root登录的密码
elasticsearch-6.0.0.rpm          100%  298     0.3KB/s   00:00
```

步骤2: 安装方法介绍

```
[root@elk-3 ~]# ls
anaconda-ks.cfg  elasticsearch-6.0.0.rpm
```

elk-1节点:

```
[root@elk-1 ~]# rpm -ivh elasticsearch-6.0.0.rpm
//参数含义：i表示安装，v表示显示安装过程，h表示显示进度
warning: elasticsearch-6.0.0.rpm: Header V4 RSA/SHA512 Signature, key ID d88e42b4: NOKEY
Preparing...                          ################################# [100%]
Creating elasticsearch group... OK
Creating elasticsearch user... OK
Updating / installing...
   1:elasticsearch-0:6.0.0-1          ################################# [100%]
### NOT starting on installation, please execute the following statements to configure elasticsearch service to start automatically using systemd
 sudo systemctl daemon-reload
 sudo systemctl enable elasticsearch.service
### You can start elasticsearch service by executing
 sudo systemctl start elasticsearch.service
```

elk-2节点:

```
[root@elk-2 ~]# rpm -ivh elasticsearch-6.0.0.rpm
warning: elasticsearch-6.0.0.rpm: Header V4 RSA/SHA512 Signature, key ID d88e42b4: NOKEY
Preparing...                          ################################# [100%]
Creating elasticsearch group... OK
Creating elasticsearch user... OK
Updating / installing...
   1:elasticsearch-0:6.0.0-1          ################################# [100%]
### NOT starting on installation, please execute the following statements to configure elasticsearch service to start automatically using systemd
 sudo systemctl daemon-reload
 sudo systemctl enable elasticsearch.service
### You can start elasticsearch service by executing
 sudo systemctl start elasticsearch.service
```

elk-3节点:

```
[root@elk-3 ~]# rpm -ivh elasticsearch-6.0.0.rpm
warning: elasticsearch-6.0.0.rpm: Header V4 RSA/SHA512 Signature, key ID
```

```
d88e42b4: NOKEY
   Preparing...                ################################# [100%]
   Creating elasticsearch group... OK
   Creating elasticsearch user... OK
Updating / installing...
   1:elasticsearch-0:6.0.0-1  ################################# [100%]
### NOT starting on installation, please execute the following statements to configure elasticsearch service to start automatically using systemd
   sudo systemctl daemon-reload
   sudo systemctl enable elasticsearch.service
### You can start elasticsearch service by executing
   sudo systemctl start elasticsearch.service
```

学习笔记

考评记录

姓名		完成日期	
序号	考核内容	标准分	评分
1	远程复制文件	50	
2	安装方法介绍	50	
	总评分	100	
任务实现心得：			

任务实训	如果 elk-3 节点查看没有复制过去应该怎么做
任务目标	了解项目中可能出错的地方

任务 3　配置 Elasticsearch

任务描述

情境描述	程序员A在下载好Elasticsearch的基础上，发现还需要配置文件才能让自己更好地使用，所以接下来需要对它进行配置
任务分解	分析上面的工作情境，将任务分解如下： （1）配置Elasticsearch的配置文件； （2）启动服务
任务准备	熟悉Linux系统的基本指令

任务目标

知识目标	了解Elasticsearch的配置文件编写和服务的启动方式
技能目标	学会启动服务的方式
素质目标	耐心与细心：在Elasticsearch的配置文件和启动过程中，针对配置文件报错和启动失败问题耐心地查找和解决问题

任务实现

步骤1： 配置文件/etc/elasticsearch/elasticsearch.yml。

（1）elk-1节点：增加以下加粗字样（//为解释，这里用不到的配置文件被删除），注意IP。

```
[root@elk-1 ~]# vi /etc/elasticsearch/elasticsearch.yml
[root@elk-1 ~]# cat /etc/elasticsearch/elasticsearch.yml
# ======= Elasticsearch Configuration ===========
#
# NOTE: Elasticsearchcomes with reasonable defaults for most settings.
#       Before you set out to tweak and tune the configuration, make sure you
#       understand what are you trying to accomplish and the consequences.
#
# The primary way of configuring a node is via this file. This template lists
# the most important settings you may want to configure for a production cluster.
#
# Please consult the documentation for further information on configuration options:
# https://www.elastic.co/guide/en/elasticsearch/reference/index.html
#
# ------------------Cluster --------------------
# Use a descriptive name for your cluster:
```

```
cluster.name: ELK
//配置es的集群名称，默认是elasticsearch，es会自动发现在同一网段下的es，如果在同一
网段下有多个集群，就可以用这个属性来区分不同的集群
# ------------------------Node ------------------
# Use a descriptive name for the node:
node.name: elk-1
//节点名，默认随机指定一个name列表中名字，该列表在es的jar包中config文件夹的name.txt
文件中，其中有很多作者添加的名字
node.master: true
//指定该节点是否有资格被选举成为node，默认是true，es是默认集群中的第一台机器为master，
如果这台机器挂了就会重新选举master。其他两节点为false
node.data: false
//指定该节点是否存储索引数据，默认为true。其他两节点为true
# ----------------- Paths ----------------
# Path to directory where to store the data (separate multiple locations by comma):
path.data: /var/lib/elasticsearch
//索引数据存储位置（保持默认，不要开启注释）
# Path to log files:
path.logs: /var/log/elasticsearch
//设置日志文件的存储路径，默认是es根目录下的logs文件夹
# --------------- Network ------------------
# Set the bind address to a specific IP (IPv4 or IPv6):
network.host: 192.168.40.11
//设置绑定的IP地址，可以是IPv4或IPv6的，默认为0.0.0.0
# Set a custom port for HTTP:
http.port: 9200
//启动的es对外访问的HTTP端口，默认为9200
# For more information, consult the network module documentation.
# --------------------Discovery ---------------
# Pass an initial list of hosts to perform discovery when new node is started:
# The default list of hosts is ["127.0.0.1", "[::1]"]
#discovery.zen.ping.unicast.hosts: ["host1", "host2"]
discovery.zen.ping.unicast.hosts: ["elk-1","elk-2","elk-3"]
//设置集群中master节点的初始列表，可以通过这些节点自动发现新加入集群的节点
```

(2) elk-2节点：

```
[root@elk-2 ~]# vi /etc/elasticsearch/elasticsearch.yml
[root@elk-2 ~]# cat /etc/elasticsearch/elasticsearch.yml |grep -v ^# |grep -v ^$
cluster.name: ELK
node.name: elk-2
node.master: false
node.data: true
path.data: /var/lib/elasticsearch
```

```
path.logs: /var/log/elasticsearch
network.host: 192.168.40.12
http.port: 9200
discovery.zen.ping.unicast.hosts: ["elk-1","elk-2","elk-3"]
```

(3) elk-3节点：

```
[root@elk-2 ~]# vi /etc/elasticsearch/elasticsearch.yml
[root@elk-2 ~]# cat /etc/elasticsearch/elasticsearch.yml |grep -v ^# |grep -v ^$
cluster.name: ELK
node.name: elk-3
node.master: false
node.data: true
path.data: /var/lib/elasticsearch
path.logs: /var/log/elasticsearch
network.host: 192.168.40.13
http.port: 9200
discovery.zen.ping.unicast.hosts: ["elk-1","elk-2","elk-3"]
```

步骤2： 启动服务。

通过命令启动es服务，启动后使用ps命令查看进程是否存在或者使用netstat命令查看是否端口启动。命令如下：（三个节点命令相同）

```
[root@elk-1 ~]# systemctl start elasticsearch
[root@elk-1 ~]# ps -ef|grepelasticsearch
elastic+  19280     1  0 09:00 ?        00:00:54 /bin/java -Xms1g
-Xmx1g -XX:+UseConcMarkSweepGC -XX:CMSInitiatingOccupancyFraction=75
-XX:+UseCMSInitiatingOccupancyOnly -XX:+AlwaysPreTouch -server -Xss1m
-Djava.awt.headless=true -Dfile.encoding=UTF-8 -Djna.nosys=true
-XX:-OmitStackTraceInFastThrow -Dio.netty.noUnsafe=true
-Dio.netty.noKeySetOptimization=true
-Dio.netty.recycler.maxCapacityPerThread=0
-Dlog4j.shutdownHookEnabled=false -Dlog4j2.disable.jmx=true
-XX:+HeapDumpOnOutOfMemoryError
-XX:HeapDumpPath=/var/lib/elasticsearch
-Des.path.home=/usr/share/elasticsearch -Des.path.conf=/etc/elasticsearch -cp
/usr/share/elasticsearch/lib/* org.elasticsearch.bootstrap.Elasticsearch -p
/var/run/elasticsearch/elasticsearch.pid --quiet
root      19844 19230  0 10:54 pts/0    00:00:00 grep --color=auto
elasticsearch
[root@elk-1 ~]# netstat -lntp
Active Internet connections (only servers)
```

Proto	Recv-Q	Send-Q	Local Address	Foreign Address	State	PID/Program name
tcp	0	0	0.0.0.0:22	0.0.0.0:*	LISTEN	1446/sshd
tcp	0	0	127.0.0.1:25	0.0.0.0:*	LISTEN	1994/master
tcp6	0	0	192.168.40.11:9200	:::*	LISTEN	19280/java
tcp6	0	0	192.168.40.11:9300	:::*	LISTEN	19280/java
tcp6	0	0	:::22	:::*	LISTEN	1446/sshd
tcp6	0	0	::1:25	:::*	LISTEN	1994/master

> **说明：**
> 有以上端口或者进程存在，证明 es 服务启动成功。

学习笔记

任务考评

考评记录

姓名			完成日期	
序号	考核内容		标准分	评分
1	配置文件		50	
2	启动服务成功		50	
	总评分		100	

任务实现心得：

任务实训

任务实训	在服务启动过程中发现服务没有完全启动的解决方案
任务目标	学会服务启动的解决方法

单元 2　检测集群状态和部署 Kibana

本单元包括两个任务：任务 1 是检测集群状态，具体指使用 curl 命令检查集群状态；任务 2 是部署 Kibana，包括安装 Kibana、配置 Kibana、启动 Kibana。

学习目标

通过本单元的学习，使学生掌握检测集群状态和部署 Kibana 的基本知识，培养学生自主进行检测集群状态和部署 Kibana 的能力。

任务 1　检测集群状态

任务描述

情境描述	程序员A在安装好Elasticsearch之后想要对服务器的集群状态进行检测，知道各个集群的运行状态，发现通过命令可以实现
任务分解	分析上面的工作情境，需要完成使用curl命令检查集群状态
任务准备	熟悉Linux系统的基本指令

任务目标

知识目标	掌握查看集群状态的方法
技能目标	学会curl命令的使用方法和原理
素质目标	耐心与细心：在curl命令的使用过程中，了解各个参数的意义，探索达到不同检测目的的不同命令

任务实现

步骤： 使用curl命令检查集群状态。

elk-1节点：

```
[root@elk-1 ~]# curl '192.168.40.11:9200/_cluster/health?pretty'
{
  "cluster_name" : "ELK",
  "status" : "green",
//为green表示集群没问题，yellow或者red则表示集群有问题
  "timed_out" : false,
```

```
    //是否有超时
    "number_of_nodes" : 3,
    //集群中的节点数量
    "number_of_data_nodes" : 2,
    //集群中data节点的数量
    "active_primary_shards" : 1,
    "active_shards" : 2,
    "relocating_shards" : 0,
    "initializing_shards" : 0,
    "unassigned_shards" : 0,
    "delayed_unassigned_shards" : 0,
    "number_of_pending_tasks" : 0,
    "number_of_in_flight_fetch" : 0,
    "task_max_waiting_in_queue_millis" : 0,
    "active_shards_percent_as_number" : 100.0
}
```

学习笔记

考评记录

姓名		完成日期	
序号	考核内容	标准分	评分
1	curl 命令检测	50	
2	改变检测目的也能使用 curl 命令查看	50	
	总评分	100	

任务实现心得：

任务实训

任务实训	学习 curl 命令实现的底层原理
任务目标	掌握 Linux 服务器的底层实现原理

任务 2　部署 Kibana

任务描述

情境描述	程序员A在查阅相关资料时发现ELK中K的全称为Kibana。据了解，Kibana是一个针对Elasticsearch的开源分析及可视化平台，用来搜索、查看交互存储在Elasticsearch索引中的数据。使用Kibana，可以通过各种图表进行高级数据分析及展示。Kibana让海量数据更容易理解。它操作简单，基于浏览器的用户界面可以快速创建仪表板（dashboard）实时显示Elasticsearch查询动态。所以在项目开展过程中，非常有必要使用到这个工具
任务分解	分析上面的工作情境，将任务分解如下： （1）安装Kibana； （2）配置Kibana； （3）启动Kibana
任务准备	熟悉Linux系统的基本指令，掌握Kibana的基本作用

任务目标

知识目标	了解Kibana的安装、配置、部署过程
技能目标	掌握下载Kibana的方法，学会安装和部署该工具
素质目标	耐心与细心：在Kibana的配置和启动过程中，针对配置报错和启动失败问题耐心地查找和解决问题

任务实现

步骤1： 安装Kibana。

通过scrt工具把kibana的rpm包上传至elk-1节点的root的目录下（其他节点不需要上传）。

```
[root@elk-1 ~]# rpm -ivh kibana-6.0.0-x86_64.rpm
warning: kibana-6.0.0-x86_64.rpm: Header V4 RSA/SHA512 Signature, key ID d88e42b4: NOKEY
Preparing...                          ################################# [100%]
Updating / installing...
   1:kibana-6.0.0-1                   ################################# [100%]
```

知识链接

Kibana是一个开源的分析与可视化平台，通常和Elasticsearch一起使用。可以用kibana搜索、查看存放在Elasticsearch中的数据。Kibana与Elasticsearch的交互方式是各种不同的图表、表格、地图等，直观地展示数据，从而达到数据分析与可视化的目的。

Elasticsearch、Logstash和Kibana这三个技术的组合是大数据领域中一个很巧妙的设计。一种很典型的MVC思想，即模型持久层、控制层和视图层。Logstash担任控制层的角色，负责搜集和过滤数据；Elasticsearch担任数据持久层的角色，负责存储数据；而这章的主题Kibana担任视图层角色，拥有各种维度的查询和分析，并使用图形化的界面展示存放在Elasticsearch中的数据。

步骤2：配置Kibana。

配置Kibana的配置文件在/etc/kibana/kibana.yml中，在其中增加或修改以下内容：

```
[root@elk-1 ~]# vi /etc/kibana/kibana.yml
[root@elk-1 ~]# cat /etc/kibana/kibana.yml|grep -v ^#
server.port: 5601
server.host: 192.168.40.11
elasticsearch.url: "http://192.168.40.11:9200"
```

步骤3：启动Kibana。

```
[root@elk-1 ~]# systemctl start kibana
[root@elk-1 ~]# ps -ef|grepkibana
kibana  19958    1 41 11:26 ?        00:00:03
/usr/share/kibana/bin/../node/bin/node --no-warnings
/usr/share/kibana/bin/../src/cli -c /etc/kibana/kibana.yml
root    19970 19230  0 11:26 pts/0 00:00:00 grep --color=auto kibana
[root@elk-1 ~]# netstat -lntp |grep node
tcp     0     0 192.168.40.11:5601     0.0.0.0:*     LISTEN     19958/node
```

启动成功后访问网页，界面如图3-2-1所示。

图 3-2-1　步骤图 1

考评记录

姓名			完成日期	
序号	考核内容		标准分	评分
1	下载 Kibana		30	
2	配置 Kibana		30	
3	启动 Kibana		40	
	总评分		100	

任务实现心得：

任务实训

任务实训	修改配置文件中的 server port 信息，查看结果是否正确
任务目标	了解配置文件中端口号的作用

单元 3　部署 Logstash 和收集 Nginx 日志

本单元包括三个任务：任务 1 是部署 Logstash，包括安装 Logstash、配置 Logstash、启动 Logstash、在 Kibana 上查看日志、Web 界面配置；任务 2 是 Logstash 收集 Nginx 日志，包括安装 Nginx、配置 Logstash、Web 界面配置；任务 3 是使用 Beats 采集日志，包括安装 Beats、配置 Beats、Web 界面配置。

学习目标

通过本单元的学习，使学生掌握 Logstash 部署和收集 Nginx 日志的基本知识，培养学生自主进行 Logstash 部署和收集 Nginx 日志的能力。

任务 1　部署 Logstash

任务描述

部署Logstash

情境描述	项目经理A在实际部署过程中发现，随着部署程序的增多，数据也在增多，因此急需部署一个对数据进行分析整合的工具。他选择了流行的数据整理分析工具Logstash
任务分解	分析上面的工作情境，将任务分解如下： （1）安装Logstash； （2）配置Logstash； （3）启动Logstash； （4）在Kibana上查看日志； （5）Web界面配置
任务准备	熟悉Linux系统的基本指令，掌握Kibana的操作方式

任务目标

知识目标	了解Logstash的使用方式
技能目标	学会Logstash的软件安装过程和使用方法
素质目标	耐心与细心：在Logstash的安装、配置和启动过程中，针对配置报错和启动失败问题耐心地查找和解决问题，对不了解的问题及时查阅相关资料解决

任务实现

步骤1: 安装Logstash。

使用scrt工具把kibana的rpm包上传至elk-2节点的root的目录下（其他节点不需上传）。

```
[root@elk-2 ~]# rpm -ivh logstash-6.0.0.0.rpm
warning: logstash-6.0.0.0.rpm: Header V4 RSA/SHA512 Signature, key ID d88e42b4: NOKEY
Preparing...                          ################################# [100%]
Updating / installing...
   1:logstash-1:6.0.0-1               ################################# [100%]
Using provided startup.options file: /etc/logstash/startup.options
```

知识链接

Logstash是一款开源的数据收集引擎，具备实时管道处理能力。简单来说，Logstash作为数据源与数据存储分析工具之间的桥梁，结合Elasticsearch以及Kibana，能够极大方便数据的处理与分析。通过200多个插件，Logstash可以接收几乎各种各样的数据，包括日志、网络请求、关系型数据库、传感器或物联网等。

步骤2: 配置Logstash。

配置/etc/logstash/logstash.yml，修改增加如下：

```
[root@elk-2 ~]# vi /etc/logstash/logstash.yml
http.host: "192.168.40.12"
```

配置logstash收集syslog日志：

```
[root@elk-2 ~]# vi /etc/logstash/conf.d/syslog.conf
[root@elk-2 ~]# cat /etc/logstash/conf.d/syslog.conf
input {     //定义日志源
    file {
        path => "/var/log/messages"      //定义日志来源路径目录要给644权限，不然
                                         //无法读取日志
        type =>"systemlog"    //定义类型
        start_position => "beginning"
        stat_interval => "3"
    }
}
output {    //定义日志输出
    if [type] == "systemlog" {
```

3-22

```
            elasticsearch {
                hosts => ["192.168.40.11:9200"]
                index => "system-log-%{+YYYY.MM.dd}"
        }
    }
}
```

检测配置文件是否错误:

```
[root@elk-2 ~]# ln -s /usr/share/logstash/bin/logstash /usr/bin
//创建软连接,方便使用logstash命令
[root@elk-2 ~]# logstash --path.settings /etc/logstash/ -f
/etc/logstash/conf.d/syslog.conf --config.test_and_exit
Sending Logstash's logs to /var/log/logstash which is now configured via
log4j2.properties
Configuration OK
//为ok则代表没问题
```

> **说明**:
> - --path.settings: 用于指定 logstash 的配置文件所在的目录。
> - -f: 指定需要被检测的配置文件的路径。
> - --config.test_and_exit: 指定检测完成之后退出,否则会直接启动。

步骤3: 启动Logstash。

检查配置文件没有问题后,启动Logstash服务:

```
[root@elk-2 ~]# systemctl start logstash
```

使用命令ps,查看进程:

```
[root@elk-2 ~]#ps -ef|grep logstash
  logstash   21835      1 12 16:45 ?        00:03:01 /bin/java
-XX:+UseParNewGC -XX:+UseConcMarkSweepGC
-XX:CMSInitiatingOccupancyFraction=75
-XX:+UseCMSInitiatingOccupancyOnly -XX:+DisableExplicitGC
-Djava.awt.headless=true -Dfile.encoding=UTF-8
-XX:+HeapDumpOnOutOfMemoryError -Xmx1g -Xms256m -Xss2048k
-Djffi.boot.library.path=/usr/share/logstash/vendor/jruby/lib/jni
-Xbootclasspath/a:/usr/share/logstash/vendor/jruby/lib/jruby.jar -classpath :
-Djruby.home=/usr/share/logstash/vendor/jruby
-Djruby.lib=/usr/share/logstash/vendor/jruby/lib
-Djruby.script=jruby -Djruby.shell=/bin/shorg.jruby.Main
```

```
/usr/share/logstash/lib/bootstrap/environment.rblogstash/runner.rb --path.settings
/etc/logstash
  root     21957  20367  0 17:10 pts/2   00:00:00 grep --color=auto logstash
```

使用netstat命令，查看进程端口：

```
[root@elk-2 ~]# netstat -lntp
Active Internet connections (only servers)
Proto Recv-Q Send-Q Local Address          Foreign Address       State      PID/Program name
tcp       0      0 0.0.0.0:22             0.0.0.0:*             LISTEN     1443/sshd
tcp       0      0 127.0.0.1:25           0.0.0.0:*             LISTEN     2009/master
tcp6      0      0 192.168.40.12:9200     :::*                  LISTEN     19365/java
tcp6      0      0 :::10514               :::*                  LISTEN     21835/java
tcp6      0      0 192.168.40.12:9300     :::*                  LISTEN     19365/java
tcp6      0      0 :::22                  :::*                  LISTEN     1443/sshd
tcp6      0      0 ::1:25                 :::*                  LISTEN     2009/master
tcp6      0      0 192.168.40.12:9600     :::*                  LISTEN     21835/java
```

可能遇到的问题：

（1）启动服务后，有进程但是没有端口。首先查看日志内容，如图3-3-1所示。

```
[root@elk-2 ~]# cat /var/log/logstash/logstash-plain.log|grep que
```

```
[root@localhost ~]# cat /var/log/logstash/logstash-plain.log | grep que
[2020-04-09T15:27:12,225][INFO ][logstash.setting.writabledirectory] Creating direc
tory {:setting=>"path.queue", :path=>"/var/lib/logstash/queue"}
[2020-04-09T15:27:12,226][INFO ][logstash.setting.writabledirectory] Creating direc
tory {:setting=>"path.dead_letter_queue", :path=>"/var/lib/logstash/dead_letter_que
ue"}
[root@localhost ~]#
```

图 3-3-1　步骤图 1

（2）通过日志确定是权限问题，因为之前以root的身份在终端启动过Logstash，所以产生的相关文件的权限用户和权限组都是root：

```
[root@elk-2 ~]# ll /var/lib/logstash/
total 4
drwxr-xr-x. 2 root root    6 Dec  6 15:45 dead_letter_queue
drwxr-xr-x. 2 root root    6 Dec  6 15:45 queue
-rw-r--r--. 1 root root   36 Dec  6 15:45 uuid
```

（3）修改/var/lib/logstash/目录的所属组为logstash，并重启服务：

```
[root@elk-2 ~]# chown -R logstash /var/lib/logstash/
[root@elk-2 ~]# systemctl restart logstash            //重启服务后即可
```

步骤4： Kibana上查看日志。

之前部署Kibana完成后，还没有检索日志。现在Logstash部署完成，回到Kibana服务器上查看日志索引，执行命令如下：

```
[root@elk-1 ~]# curl '192.168.40.11:9200/_cat/indices?v'
health status index                     uuid           pri rep
docs.count docs.deleted store.size pri.store.size
green   open   system-log-2019.12.06  UeKk3IY6TiebNu_OD04YZA    5
1        938           0         816kb         412.2kb
green   open   .kibana                KL7WlNw_T7K36_HSbchBcw
1        1     1       0         7.3kb         3.6kb
```

获取/删除指定索引详细信息：

```
[root@elk-1 ~]# curl -XGET/DELETE '192.168.40.11:9200/system-log-2020.04.09?pretty'

{
  "system-log-2020.04.09" : {
    "aliases" : { },
    "mappings" : {
      "systemlog" : {
        "properties" : {
          "@timestamp" : {
            "type" : "date"
          },
          "@version" : {
            "type" : "text",
            "fields" : {
              "keyword" : {
                "type" : "keyword",
                "ignore_above" : 256
              }
            }
          },
          "host" : {
            "type" : "text",
            "fields" : {
              "keyword" : {
                "type" : "keyword",
                "ignore_above" : 256
              }
            }
          },
          "message" : {
```

```
          "type" : "text",
          "fields" : {
            "keyword" : {
              "type" : "keyword",
              "ignore_above" : 256
            }
          }
        },
        "path" : {
          "type" : "text",
          "fields" : {
            "keyword" : {
              "type" : "keyword",
              "ignore_above" : 256
            }
          }
        },
        "type" : {
          "type" : "text",
          "fields" : {
            "keyword" : {
              "type" : "keyword",
              "ignore_above" : 256
            }
          }
        }
      }
    },
    "settings" : {
      "index" : {
        "creation_date" : "1586417513863",
        "number_of_shards" : "5",
        "number_of_replicas" : "1",
        "uuid" : "ETzosODmRA2bksRXxIiabQ",
        "version" : {
          "created" : "6000099"
        },
        "provided_name" : "system-log-2020.04.09"
      }
    }
  }
}
```

步骤5: Web界面配置。

浏览器访问192.168.40.11:5601，到Kibana上配置索引，如图3-3-2所示。

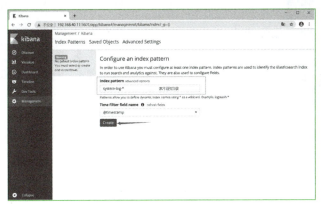

图 3-3-2　步骤图 2

配置完成后，选择Discover，进入Discover页面后，如果出现图3-3-3所示提示，则代表无法查找到日志信息。

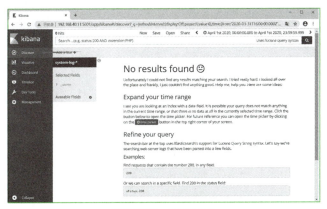

图 3-3-3　步骤图 3

这种情况一般是时间的问题，单击右上角信息切换成查看当天的日志信息。由于虚拟机的时间是2020年4月9日，所以要把时间调整到那一天，如图3-3-4所示。

图 3-3-4　步骤图 4

设置完成后就正常了，如图3-3-5所示。

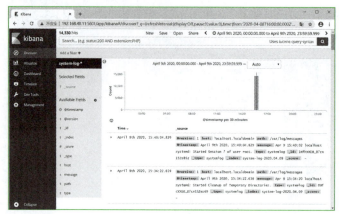

图 3-3-5　步骤图 5

任务考评

考评记录

姓名		完成日期	
序号	考核内容	标准分	评分
1	下载 Logstash	20	
2	配置 Logstash	20	
3	启动 Logstash	20	
4	在 Kibana 上查看日志	20	
5	Web 配置	20	
	总评分	100	

任务实现心得：

任务实训

任务实训	查看 Kibana 日志记录，解决 Logstash 出现的问题
任务目标	学习通过日志文件解决程序问题

任务 2　Logstash 收集 Nginx 日志

📋 任务描述

视频
Logstash收集Nginx日志

情境描述	项目经理A安装好了Logstash之后就要使用它去分析数据，首先需要找到关键数据，Nginx服务器的数据是首先要进行收集分析的，因此他就要开始收集Nginx日志了
任务分解	分析上面的工作情境，将任务分解如下： （1）安装Nginx； （2）配置Logstash； （3）Web界面配置
任务准备	熟悉Linux系统的基本指令，对Nginx服务器有基本了解。知道日志文件的存放地址与方式

🎯 任务目标

知识目标	掌握Logstash收集工具的使用方式
技能目标	掌握Logstash与Nginx服务器的日志文件联合使用与Web界面的配置
素质目标	耐心与细心：在Logstash与Nginx日志文件结合收集中，针对日志文件的相关问题耐心地查找和解决问题，对待Web配置出现的问题及时查阅相关资料解决

🛠 任务实现

步骤1： 安装Nginx。

使用wget命令下载Nginx的rpm包并安装：

```
[root@elk-2 ~]#wget
http://nginx.org/packages/centos/7/x86_64/RPMS/nginx-1.16.1-1.el7.ngx.x86_64.rpm
[root@elk-2 ~]# rpm -ivh nginx-1.16.1-1.el7.ngx.x86_64.rpm
警告：nginx-1.16.1-1.el7.ngx.x86_64.rpm: 头V4 RSA/SHA1 Signature, 密钥 ID 7bd9bf62: NOKEY
准备中...
################################ [100%]
正在升级/安装...
   1:nginx-1:1.16.1-1.el7.ngx
(  3%################################ [100%]
----------------------------------------------------------------------

Thanks for using nginx!

Please find the official documentation for nginx here:
```

```
* http://nginx.org/en/docs/

Please subscribe to nginx-announce mailing list to get
the most important news about nginx:
*http://nginx.org/en/support.html

Commercial subscriptions for nginx are available on:
*http://nginx.com/products/

--------------------------------------------------------------------
[root@elk-2 ~]#
```

知识链接

　　Nginx日志主要分为访问日志和错误日志两种。日志开关在Nginx配置文件（/etc/nginx/nginx.conf）中设置，两种日志都可以选择性关闭，默认都是打开的。通过访问日志，可以得到用户地域来源、跳转来源、使用终端、某个URL访问量等相关信息；通过错误日志，可以得到系统某个服务或Server的性能瓶颈等。因此，好好利用日志可以得到很多有价值的信息。

步骤2： 配置Logstash。

在elk-2上，编辑Nginx配置文件，加入如下内容：

```
[root@elk-2 ~]# vi /etc/logstash/conf.d/nginx.conf
input {
   file {
        path => "/tmp/elk_access.log"
        start_position => "beginning"
        type => "nginx"
   }
}
filter {
   grok {
           match => { "message" => "%{IPORHOST:http_host}%{IPORHOST:clientip}-%{USERNAME:remote_user}\[%{HTTPDATE:timestamp}\] \"(?:%{WORD:http_verb}%{NOTSPACE:http_request}(?: HTTP/%{NUMBER:http_version})?|%{DATA:raw_http_request})\" %{NUMBER:response} (?:%{NUMBER:bytes_read}|-)%{QS:referrer} %{QS:agent} %{QS:xforwardedfor}%{NUMBER:request_time:float}"}
      }
      geoip {
        source => "clientip"
```

```
        }
    }
    output {
        stdout { codec =>rubydebug }
        elasticsearch {
            hosts => ["192.168.40.12:9200"]
            index => "nginx-test-%{+YYYY.MM.dd}"
        }
    }
```

使用logstash命令检查文件是否错误:

```
[root@elk-2 ~]# logstash --path.settings /etc/logstash/ -f
/etc/logstash/conf.d/nginx.conf --config.test_and_exit
  Sending Logstash's logs to /var/log/logstash which is now configured via
log4j2.properties
  Configuration OK
```

编辑监听Nginx日志配置文件,加入如下内容:

```
[root@elk-2 ~]# vi /etc/nginx/conf.d/elk.conf
server {
    listen 80;
    server_name elk.com;

    location / {
        proxy_pass         http://192.168.40.11:5601;
        proxy_set_header Host      $host;
        proxy_set_header X-Real-IP      $remote_addr;
        proxy_set_header X-Forwarded-For $proxy_add_x_forwarded_for;
    }
    access_log  /tmp/elk_access.log main2;
}
```

修改Nginx日志配置文件,增加如下内容(需注意Nginx配置文件格式):

```
[root@elk-2 ~]# vim /etc/nginx/nginx.conf
  log_format main2'$http_host $remote_addr - $remote_user [$time_local]
"$request"' '$status $body_bytes_sent "$http_referer"'
                '"$http_user_agent" "$upstream_addr" $request_time';
[root@elk-2 ~]# nginx -t
nginx: the configuration file /etc/nginx/nginx.conf syntax is ok
nginx: configuration file /etc/nginx/nginx.conf test is successful
```

```
[root@elk-2 ~]#
[root@elk-2 ~]# systemctl start nginx
```

在/etc/hosts文件中添加如下信息：

```
192.168.40.12 elk.com
```

使用浏览器访问，检查是否有日志产生。

步骤3： Web界面配置。

使用浏览器访问192.168.40.11:5601，到Kibana上配置索引，如图3-3-6~图3-3-8所示。

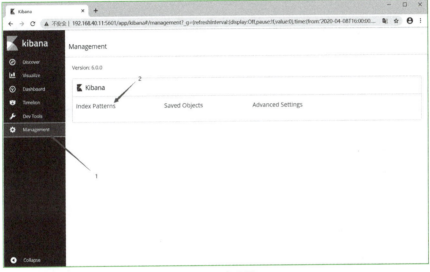

图 3-3-6　步骤图 6

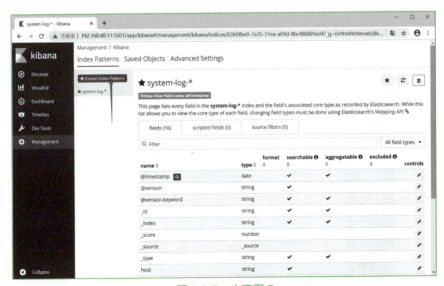

图 3-3-7　步骤图 7

图 3-3-8　步骤图 8

配置完成后，选择Discover，进入Discover页面后，如果出现图3-3-9所示提示，则代表无法查找到日志信息。

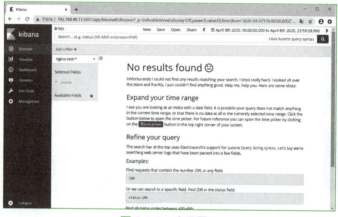

图 3-3-9　步骤图 9

这种情况一般是时间的问题，单击右上角内容切换成查看当天的日志信息。由于虚拟机的时间是2020年4月9日，所以要把时间调整到那一天，如图3-3-10和图3-3-11所示。

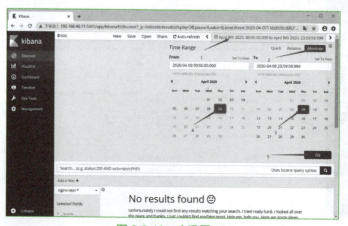

图 3-3-10　步骤图 10

项目③ ELK 日志分析系统

图 3-3-11　步骤图 11

学习笔记

考评记录

姓名			完成日期	
序号	考核内容		标准分	评分
1	下载 Nginx		30	
2	配置 Logstash		30	
3	Web 界面配置		40	
	总评分		100	

任务实现心得：

任务实训

任务实训	修改配置文件中的端口号，查看会出现的问题
任务目标	了解配置文件中端口号的作用

任务 3 使用 Beats 采集日志

任务描述

视频

使用Beats采集日志

情境描述	项目经理A发现除了之前安装好的Logstash之外，还可以通过Beats采集日志。Beats平台集合了多种单一用途数据采集器。这些采集器安装后可用作轻量型代理，从成百上千或成千上万台机器向 Logstash 或 Elasticsearch 发送数据。Beats只是用来优化Logstash的，因为Logstash消耗的性能比较多。如果只是单纯地为了收集日志，使用Logstash就有点大材小用了，还有点浪费资源，此时可以使用Beats
任务分解	分析上面的工作情境，将任务分解如下： (1) 安装Beats； (2) 配置Beats； (3) Web界面配置
任务准备	熟悉Linux系统的基本指令，对Logstash和Beats软件的优缺点有所了解

任务目标

知识目标	掌握Beats收集工具的使用方式
技能目标	学会Beats的安装和配置过程，了解使用Beats的优点
素质目标	耐心与细心：针对Beats工具安装和配置过程中出现的相关问题耐心地查找和解决，对Web配置中出现的问题及时查阅相关资料解决

任务实现

步骤1： 安装Beats。

在elk-3主机上下载和安装Beats：

```
[root@elk-3 ~]# wget
https://artifacts.elastic.co/downloads/beats/filebeat/filebeat-6.0.0-x86_64.rpm
 --2020-03-30 22:41:52--
https://artifacts.elastic.co/downloads/beats/filebeat/filebeat-6.0.0-x86_64.rpm
 正在解析主机 artifacts.elastic.co (artifacts.elastic.co)... 151.101.230.222, 2a04:4e42:1a::734
 正在连接 artifacts.elastic.co (artifacts.elastic.co)|151.101.230.222|:443... 已连接。
 已发出 HTTP 请求，正在等待回应... 200 OK
 长度: 11988378 (11M) [binary/octet-stream]
 正在保存至: "filebeat-6.0.0-x86_64.rpm.1"

100%[===========================================>] 11,988,378   390KB/s  用时 30s
```

```
2020-03-30 22:42:24 (387 KB/s) - 已保存 "filebeat-6.0.0-x86_64.rpm.1"
[11988378/11988378])

[root@elk-3 ~]# rpm -ivh  filebeat-6.0.0-x86_64.rpm
```

步骤2： 配置Beats。

编辑配置文件，如图3-3-12和图3-3-13所示。

```
[root@elk-3 ~]# vim /etc/filebeat/filebeat.yml
filebeat.prospectors:
paths:
    - /var/log/elasticsearch/elk.log         //此处可自行改为想要监听的日志文件
output.elasticsearch:
hosts: ["192.168.40.11:9200"]
systemctl start   filebeat
```

图 3-3-12　步骤图 12

图 3-3-13　步骤图 13

在elk-1主机上使用curl '192.168.40.11:9200/_cat/indices?v'命令查看是否监听到elk-3主机上的日志（出现filebeat字样表示成功），如图3-3-14所示。

图 3-3-14　步骤图 14

步骤3: Web界面配置。

采用同样的方法，在浏览器中添加filebeat日志文件的监控，如图3-3-15和图3-3-16所示。

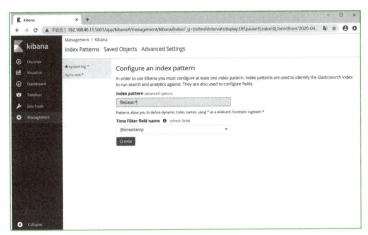

图 3-3-15　步骤图 15

图 3-3-16　步骤图 16

学习笔记

考评记录

姓名			完成日期	
序号	考核内容		标准分	评分
1	下载 Beats		30	
2	配置 Beats		30	
3	Web 界面配置		40	
	总评分		100	

任务实现心得：

任务实训	修改配置文件中想要监听的文件名，看看会发生什么
任务目标	掌握配置文件中监听文件地址的含义

项目 4
MySQL 常用操作

本项目主要介绍 MySQL 常用操作，共分为四个单元，分别介绍了更改 root 密码与连接 MySQL、MySQL 的常用操作与创建用户、MySQL 中常用 SQL 语句、MySQL 主从配置。

云计算应用运维实战

单元 1　更改 root 密码与连接 MySQL

视频

MySQL常用操作及更改root密码

本单元包括两个任务：任务 1 是更改 root 密码，包括更改环境变量、创建 MySQL 密码、密码重置；任务 2 是连接 MySQL，需要使用四条命令来使得服务器连接 MySQL 数据库。

学习目标

通过本单元的学习，使学生掌握更改 root 密码与连接 MySQL 的基本知识，培养学生自主进行更改 root 密码与连接 MySQL 的能力。

任务 1　更改 root 密码

任务描述

情境描述	程序员A接到任务连接MySQL数据库，发现root密码错误连接不上，所以需要更改root密码
任务分解	分析上面的工作情境，将任务分解如下： （1）更改环境变量； （2）创建MySQL密码； （3）密码重置
任务准备	了解Linux操作和MySQL数据库的基本知识

任务目标

知识目标	了解root密码与MySQL数据库之间的关系
技能目标	学会如何修改root密码
素质目标	耐心与细心：针对修改root密码带来的权限问题和报错问题进行细心的分析，MySQL数据库的相关指令与配置能够耐心地完成

任务实现

步骤1：更改环境变量。

修改/etc/profile文件，在文件结尾添加MySQL的绝对路径，如图4-1-1所示。

图 4-1-1　修改环境变量

4-2

步骤2：创建MySQL密码。

使用mysqladmin -uroot password '123456'命令为root用户创建初始密码，如图4-1-2所示。

```
[root@localhost ~]# mysqladmin -uroot password '123456'
Warning: Using a password on the command line interface can be insecure.
```

图 4-1-2　创建初始密码

注释：可以忽略 warning 内容，其指的是明码输入屏幕不安全。

使用mysql -uroot -p123456命令，完成初始密码登录，如图4-1-3所示。

```
[root@localhost ~]# mysql -uroot -p123456
Warning: Using a password on the command line interface can be insecure.
Welcome to the MySQL monitor.  Commands end with ; or \g.
Your MySQL connection id is 3
Server version: 5.6.43 MySQL Community Server (GPL)

Copyright (c) 2000, 2019, Oracle and/or its affiliates. All rights reserved.

Oracle is a registered trademark of Oracle Corporation and/or its
affiliates. Other names may be trademarks of their respective
owners.

Type 'help;' or '\h' for help. Type '\c' to clear the current input statemen
mysql>
```

图 4-1-3　使用初始密码完成登录

步骤3：密码重置。

修改配置文件/etc/my.cnf，在mysqld配置段增加字段skip-grant，如图4-1-4所示。

```
[mysqld]
skip-grant
# Remove leading # and set to the amount of RAM for the most important data
# cache in MySQL. Start at 70% of total RAM for dedicated server, else 10%.
# innodb_buffer_pool_size = 128M

# Remove leading # to turn on a very important data integrity option: logging
# changes to the binary log between backups.
# log_bin
```

图 4-1-4　修改配置文件配置段

修改完成后，重启MySQL服务：

```
/etc/init.d/mysqld restart
```

使用命令登录MySQL（修改的配置段是完成忽略授权的操作，可以直接登录，无须输入用户名和密码），切换到MySQL库，对user表进行更新操作，如图4-1-5所示。

```
mysql> use mysql;
Reading table information for completion of table and column names
You can turn off this feature to get a quicker startup with -A

Database changed
mysql> update user set password=password('aminglinux') where user='root';
Query OK, 4 rows affected (0.00 sec)
```

图 4-1-5　重置命令

修改完成后，确认新密码登录有效。把/etc/my.cnf改回原有状态，并重启mysql服务。

任务考评

考评记录

姓名			完成日期	
序号	考核内容		标准分	评分
1	更改环境变量		30	
2	创建 MySQL 密码		30	
3	密码重置		40	
	总评分		100	

任务实现心得：

任务实训

任务实训	登录 MySQL 创建新的库表，进行基本操作
任务目标	学会部分 SQL 语句

项目 4　MySQL 常用操作

任务 2　连接 MySQL

📄 任务描述

连接MySQL

情境描述	程序员A在更改完root密码之后需要连接MySQL数据库才能进行后面的项目，在网上查阅了相关资料如何使用服务器连接MySQL数据库
任务分解	分析上面的工作情境，将任务分解如下： （1）执行"mysql -uroot -p123456"命令； （2）执行"mysql -uroot -p123456 -h127.0.0.1 -P3306"命令； （3）执行"mysql -uroot -p123456 -S/tmp/mysql.sock"命令； （4）执行"mysql -uroot -p123456 -e "show databases""命令
任务准备	了解Linux操作和MySQL数据库的基本知识

📋 任务目标

知识目标	了解连接MySQL的方式
技能目标	学会四条命令使服务器连接MySQL数据库
素质目标	耐心与细心：针对输入指令的报错问题进行细心的分析，MySQL数据库的相关指令与配置能够耐心地完成

📝 任务实现

步骤1： 执行"mysql -uroot -p123456"命令，如图4-1-6所示。

图 4-1-6　步骤 1

步骤2： 执行"mysql -uroot -p123456 -h127.0.0.1 -P3306"命令，如图4-1-7所示。

图 4-1-7　步骤 2

步骤3: 执行"mysql -uroot -p123456 -S/tmp/mysql.sock"命令,如图4-1-8所示。

图 4-1-8　步骤 3

步骤4: 执行"mysql -uroot -p123456 -e "show databases""命令,如图4-1-9所示。

图 4-1-9　步骤 4

知识链接

MySQL 是一个关系型数据库管理系统,由瑞典 MySQL AB 公司开发,目前属于 Oracle 旗下公司。MySQL 是流行的关系型数据库管理系统,在 Web 应用方面是最好的 RDBMS (Relational Database Management System,关系数据库管理系统)应用软件之一。

数据库(Database)是按照数据结构来组织、存储和管理数据的仓库。

每个数据库都有一个或多个不同的 API 用于创建、访问、管理、搜索和复制所保存的数据。

也可以将数据存储在文件中,但是在文件中读写数据的速度相对较慢。

所以,可以使用关系型数据库管理系统(RDBMS)来存储和管理大数据量。关系型数据库是建立在关系模型基础上的数据库,借助于集合代数等数学概念和方法来处理数据库中的数据。

RDBMS 的特点如下:

(1)数据以表格的形式出现;

(2)每行为各种记录名称;

(3)每列为记录名称所对应的数据域;

(4)许多行和列组成一张表单;

(5)若干表单组成 database。

考评记录

姓名		完成日期	
序号	考核内容	标准分	评分
1	执行"mysql -uroot -p123456"命令	25	
2	执行"mysql -uroot -p123456 -h127.0.0.1 -P3306"命令	25	
3	执行"mysql -uroot -p123456 -S/tmp/mysql.sock"命令	25	
4	执行"mysql -uroot -p123456 -e "show databases""命令	25	
	总评分	100	

任务实现心得：

任务实训

任务实训	熟悉这四条指令每个参数的意义
任务目标	加深指令的底层逻辑理解

单元 2　MySQL 的常用操作与创建用户以及授权

本单元包括两个任务：任务 1 是 MySQL 的常用操作，包括查询库"show databases;"、切换库"use mysql;"、查看库中的表"show tables;"、查看表中的字段"desc user;"；任务 2 是创建用户以及授权，包括进行授权以及查看授权表。

学习目标

通过本单元的学习，使学生掌握 MySQL 的常用操作与创建用户以及授权的基本知识，培养学生自主进行 MySQL 常用操作与创建用户以及授权的能力。

任务 1　MySQL 的常用操作

任务描述

情境描述	项目经理 B 在对项目的审核过程中发现，目前服务器已经与 MySQL 数据库进行连接，但对于 MySQL 数据库的基本操作还有些不太了解，这个瓶颈问题会导致后续项目出现困难，所以这个问题急需解决
任务分解	分析上面的工作情境，将任务分解如下： (1) 执行查询库命令 "show databases;"； (2) 执行切换库命令 "use mysql;"； (3) 执行查看库中的表命令 "show tables;"； (4) 执行查看表中的字段命令 "desc user;"； (5) 执行查看建表语句命令 "show create table user\G;"； (6) 执行查看当前用户命令 "select user();"； (7) 执行查看当前使用的数据库命令 "select database();"； (8) 执行创建库命令 "create database db1;"； (9) 执行创建表命令 "use db1; create table t1('id' int(4), 'name' char(40));"； (10) 执行查看当前数据库版本命令 "select version();"； (11) 执行查看数据库状态命令 "show status;"； (12) 执行查看各参数命令 "show variables; show variables like 'max_connect%';"； (13) 执行修改参数命令 "set global max_connect_errors=1000;"； (14) 执行查看队列命令 "show processlist; show full processlist;"
任务准备	服务器可以连接 MySQL 数据库，了解数据库的相关知识

任务目标

知识目标	掌握 MySQL 数据库的常用操作
技能目标	能够熟练使用步骤中的详细指令
素质目标	耐心与细心：针对输入指令的报错问题进行细心的分析，MySQL 数据库的相关指令与配置能够耐心地完成

项目 4　MySQL 常用操作

视　频

MySQL常用命令

任务实现

步骤1： 执行查询库命令"show databases;"，如图4-2-1所示。

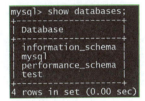

图 4-2-1　查询数据库

步骤2： 执行切换库命令"use mysql;"，如图4-2-2所示。

```
mysql> use mysql;
Reading table information for completion of table and column names
You can turn off this feature to get a quicker startup with -A

Database changed
```

图 4-2-2　切换库

步骤3： 执行查看库中的表命令"show tables;"，如图4-2-3所示。

图 4-2-3　查看库中的表

步骤4： 执行查看user表中的字段命令"desc user;"，如图4-2-4所示。

图 4-2-4　查看表中的字段

4-9

步骤5： 执行查看建表语句命令“show create table user\G;”，如图4-2-5所示。

```
mysql> show create table user\G;
*************************** 1. row ***************************
       Table: user
Create Table: CREATE TABLE `user` (
  `Host` char(60) COLLATE utf8_bin NOT NULL DEFAULT '',
  `User` char(16) COLLATE utf8_bin NOT NULL DEFAULT '',
  `Password` char(41) CHARACTER SET latin1 COLLATE latin1_bin NOT NULL DEFAULT '',
  `Select_priv` enum('N','Y') CHARACTER SET utf8 NOT NULL DEFAULT 'N',
  `Insert_priv` enum('N','Y') CHARACTER SET utf8 NOT NULL DEFAULT 'N',
  `Update_priv` enum('N','Y') CHARACTER SET utf8 NOT NULL DEFAULT 'N',
  `Delete_priv` enum('N','Y') CHARACTER SET utf8 NOT NULL DEFAULT 'N',
  `Create_priv` enum('N','Y') CHARACTER SET utf8 NOT NULL DEFAULT 'N',
  `Drop_priv` enum('N','Y') CHARACTER SET utf8 NOT NULL DEFAULT 'N',
  `Reload_priv` enum('N','Y') CHARACTER SET utf8 NOT NULL DEFAULT 'N',
  `Shutdown_priv` enum('N','Y') CHARACTER SET utf8 NOT NULL DEFAULT 'N',
  `Process_priv` enum('N','Y') CHARACTER SET utf8 NOT NULL DEFAULT 'N',
  `File_priv` enum('N','Y') CHARACTER SET utf8 NOT NULL DEFAULT 'N',
  `Grant_priv` enum('N','Y') CHARACTER SET utf8 NOT NULL DEFAULT 'N',
  `References_priv` enum('N','Y') CHARACTER SET utf8 NOT NULL DEFAULT 'N',
  `Index_priv` enum('N','Y') CHARACTER SET utf8 NOT NULL DEFAULT 'N',
  `Alter_priv` enum('N','Y') CHARACTER SET utf8 NOT NULL DEFAULT 'N',
  `Show_db_priv` enum('N','Y') CHARACTER SET utf8 NOT NULL DEFAULT 'N',
  `Super_priv` enum('N','Y') CHARACTER SET utf8 NOT NULL DEFAULT 'N',
  `Create_tmp_table_priv` enum('N','Y') CHARACTER SET utf8 NOT NULL DEFAULT 'N',
  `Lock_tables_priv` enum('N','Y') CHARACTER SET utf8 NOT NULL DEFAULT 'N',
  `Execute_priv` enum('N','Y') CHARACTER SET utf8 NOT NULL DEFAULT 'N',
  `Repl_slave_priv` enum('N','Y') CHARACTER SET utf8 NOT NULL DEFAULT 'N',
  `Repl_client_priv` enum('N','Y') CHARACTER SET utf8 NOT NULL DEFAULT 'N',
```

图 4-2-5　查看建表语句

步骤6： 执行查看当前用户命令“select user();”，如图4-2-6所示。

```
mysql> select user();
+---------+
| user()  |
+---------+
| root@   |
+---------+
1 row in set (0.00 sec)
```

图 4-2-6　查看当前用户

步骤7： 执行查看当前使用的数据库命令“select database();”，如图4-2-7所示。

```
mysql> select database();
+------------+
| database() |
+------------+
| NULL       |
+------------+
1 row in set (0.00 sec)
```

图 4-2-7　查看当前使用的数据库

步骤8： 执行创建库命令“create database db1;”，如图4-2-8所示。

```
mysql> create database db1;
Query OK, 1 row affected (0.00 sec)
```

图 4-2-8　创建库

步骤9： 执行创建表命令“use db1; create table t1('id' int(4), 'name' char(40));”，如图4-2-9所示。

```
mysql> use db1; create table t1(`id` int(4), `name` char(40));
Database changed
Query OK, 0 rows affected (0.06 sec)
```

图 4-2-9　创建表

步骤10： 执行查看当前数据库版本命令“select version();”，如图4-2-10所示。

```
mysql> select version();
+-----------+
| version() |
+-----------+
| 5.6.43    |
+-----------+
1 row in set (0.00 sec)
```

图 4-2-10　查看当前数据库版本

步骤11： 执行查看数据库状态命令“show status;”，如图4-2-11所示。

```
mysql> show status;
+------------------------------------------+-------------+
| Variable_name                            | Value       |
+------------------------------------------+-------------+
| Aborted_clients                          | 0           |
| Aborted_connects                         | 5           |
| Binlog_cache_disk_use                    | 0           |
| Binlog_cache_use                         | 0           |
| Binlog_stmt_cache_disk_use               | 0           |
| Binlog_stmt_cache_use                    | 0           |
| Bytes_received                           | 450         |
| Bytes_sent                               | 770         |
| Com_admin_commands                       | 0           |
| Com_assign_to_keycache                   | 0           |
| Com_alter_db                             | 0           |
| Com_alter_db_upgrade                     | 0           |
| Com_alter_event                          | 0           |
| Com_alter_function                       | 0           |
| Com_alter_procedure                      | 0           |
| Com_alter_server                         | 0           |
| Com_alter_table                          | 0           |
| Com_alter_tablespace                     | 0           |
```

图 4-2-11　查看数据库状态

步骤12： 执行查看各参数命令“show variables; show variables like 'max_connect%';”，结果如图4-2-12所示。

```
454 rows in set (0.00 sec)

+--------------------+-------+
| Variable_name      | Value |
+--------------------+-------+
| max_connect_errors | 100   |
| max_connections    | 151   |
+--------------------+-------+
2 rows in set (0.00 sec)
```

图 4-2-12　查看各参数

步骤13： 执行修改参数命令“set global max_connect_errors=1000;”，如图4-2-13所示。

```
mysql> set global max_connect_errors=1000;
Query OK, 0 rows affected (0.00 sec)
```

图 4-2-13　修改参数

步骤14： 执行查看队列命令“show processlist; show full processlist;”，如图4-2-14所示。

```
mysql> show processlist; show full processlist;
+----+------+-----------+------+---------+------+-------+------------------+
| Id | User | Host      | db   | Command | Time | State | Info             |
+----+------+-----------+------+---------+------+-------+------------------+
| 13 | root | localhost | db1  | Query   |    0 | init  | show processlist |
+----+------+-----------+------+---------+------+-------+------------------+
1 row in set (0.00 sec)

+----+------+-----------+------+---------+------+-------+-----------------------+
| Id | User | Host      | db   | Command | Time | State | Info                  |
+----+------+-----------+------+---------+------+-------+-----------------------+
| 13 | root | localhost | db1  | Query   |    0 | init  | show full processlist |
+----+------+-----------+------+---------+------+-------+-----------------------+
1 row in set (0.00 sec)
```

图 4-2-14　查看队列

任务考评

考评记录

姓名			完成日期	
序号	考核内容		标准分	评分
1	执行查询库命令"show databases;"		10	
2	执行切换库命令"use mysql;"		10	
3	执行查看库中的表命令"show tables;"		10	
4	执行查看表中的字段命令"desc user;"		5	
5	执行查看建表语句命令"show create table user\G;"		5	
6	执行查看当前用户命令"select user();"		5	
7	执行查看当前使用的数据库命令"select database();"		10	
8	执行创建库命令"create database db1;"		10	
9	执行创建表命令"use db1; create table t1('id' int(4), 'name' char(40));"		10	
10	执行查看当前数据库版本命令"select version();"		5	
11	执行查看数据库状态命令"show status;"		5	
12	执行查看各参数命令"show variables; show variables like 'max_connect%';"		5	
13	执行修改参数命令"set global max_connect_errors=1000;"		5	
14	执行查看队列命令"show processlist; show full processlist;"		5	
	总评分		100	

任务实现心得：

任务实训

任务实训	创建一个名为 test 的数据库，在其中创建名为 test1 的数据库表
任务目标	掌握 MySQL 的基本操作

任务 2　MySQL 创建用户以及授权

任务描述

MySQL创建用户以及授权

情境描述	程序员A掌握了MySQL数据库的基本操作，但是还需解决服务器严格的权限管理问题，并且随着项目参与人员越来越多，非常有必要对工作人员的权限进行管理，这样也会使项目更加安全
任务分解	分析上面的工作情境，将任务分解如下： (1) 进行授权； (2) 查看授权表
任务准备	了解Linux的权限管理和MySQL数据库的基本操作

任务目标

知识目标	了解MySQL的创建用户和授权管理
技能目标	学会创建用户和授权表的基本命令
素质目标	耐心与细心：针对输入指令的报错问题进行细心的分析，MySQL数据库的相关指令能够耐心地完成

任务实现

步骤1： 进行授权，如图4-2-15所示。

```
grant all on *.* to 'user1' identified by 'passwd';
grant SELECT,UPDATE,INSERT on db1.* to 'user2'@'127.0.0.1' identified by 'passwd';
grant all on db1.* to'user3'@'%' identified by 'passwd';
```

```
mysql> grant all on *.* to 'user1' identified by 'passwd';
Query OK, 0 rows affected (0.16 sec)

mysql> grant SELECT,UPDATE,INSERT on db1.* to 'user2'@'127.0.0.1' identified by 'passwd';
Query OK, 0 rows affected (0.03 sec)

mysql> grant all on db1.* to 'user3'@'%' identified by 'passwd';
Query OK, 0 rows affected (0.00 sec)
```

图 4-2-15　进行授权

步骤2： 查看授权表，如图4-2-16和图4-2-17所示。

```
show grants;
```

图 4-2-16　查看授权表步骤 1

```
show grants for user2@127.0.0.1;
```

图 4-2-17　查看授权表步骤 2

学习笔记

考评记录

姓名		完成日期	
序号	考核内容	标准分	评分
1	进行授权	50	
2	查看授权表	50	
	总评分	100	
任务实现心得：			

任务实训	对同项目的人员进行授权管理
任务目标	掌握 MySQL 的授权管理

单元 3 MySQL 常用 SQL 语句以及 MySQL 数据库的备份与恢复

本单元包括两个任务：任务 1 是熟悉并掌握 MySQL 常用 SQL 语句，包括查看表内行数 "select count(*) from mysql.user"，查看 db 表内的内容 "select * from mysql.db"，查看 db 表内含有 db 字段的内容 "select db from mysql.db"，搜索查看多个字段 "select db,user from mysql.db"，查询 host 为 127.0 的内容 "select * from mysql.db where host like '127.0.%'\G"，向 db1.t1 中插入内容 "insert into db1.t1 values (1, 'abc')"，把 id=1 的字段内容更新成 aaa "update db1.t1 set name='aaa' where id=1"；清空 db1.t1 表内的内容 truncate table db1.t1"，删除 db1.t1 表内的内容 "drop table db1.t1"；任务 2 是 MySQL 数据库的备份与恢复，包括备份库、恢复库、备份表、恢复表、备份所有库、只备份表结构。

● 视频

MySQL 常用SQL语句

通过本单元的学习，使学生掌握 MySQL 常用 SQL 语句以及 MySQL 数据库备份与恢复的基本知识，培养学生自主熟练使用 MySQL 常用 SQL 语句以及进行 MySQL 数据库备份与恢复的能力。

任务 1 MySQL 常用 SQL 语句

任务描述

情境描述	数据组的人已经将数据导入MySQL数据库，交接给了程序员A，让他试一试是否能够正常使用，那么，该怎样使用和测试数据库中的数据呢？SQL语句可以解决这个问题
任务分解	分析上面的工作情境，将任务分解如下： （1）查看表内行数 "select count(*) from mysql.user;"； （2）查看db表内的内容 "select * from mysql.db;"； （3）查看db表内含有db字段的内容 "select db from mysql.db;"； （4）搜索查看多个字段 "select db,user from mysql.db;"； （5）查询host为127.0的内容 "select * from mysql.db where host like '127.0.%' \ G;"； （6）向db1.t1中插入内容 "insert into db1.t1 values (1, 'abc');"； （7）把id=1的字段内容更新成aaa "update db1.t1 set name='aaa' where id=1;"； （8）清空db1.t1表内的内容 "truncate table db1.t1;"； （9）删除db1.t1表内的内容 "drop table db1.t1;"； （10）清空db1.t1数据库 "drop database db1;"
任务准备	掌握MySQL的常用指令

任务目标

知识目标	学习SQL语句
技能目标	学会表和库的基本SQL指令
素质目标	耐心与细心：针对输入指令的报错问题进行细心的分析，MySQL数据库的相关指令能够耐心地完成

任务实现

步骤1： 查看表内行数“select count(*) from mysql.user;”，如图4-3-1所示。

```
mysql> select count(*) from mysql.user;
+----------+
| count(*) |
+----------+
|        9 |
+----------+
1 row in set (0.00 sec)
```

图 4-3-1　查看表内行数

步骤2： 查看db表内的内容“select * from mysql.db;”，如图4-3-2所示。

```
mysql> select * from mysql.db;
+-----------+---------+-------+-------------+-------------+-------------+-------------+-------------+---
--------+------------+-----------------+------------+------------+-----------------------+--------------
---+------------------+----------------+---------------------+--------------------+--------------+-----
------+--------------+
| Host      | Db      | User  | Select_priv | Insert_priv | Update_priv | Delete_priv | Create_priv | Dr
op_priv | Grant_priv | References_priv | Index_priv | Alter_priv | Create_tmp_table_priv | Lock_tables_p
riv | Create_view_priv | Show_view_priv | Create_routine_priv | Alter_routine_priv | Execute_priv | Even
t_priv | Trigger_priv |
+-----------+---------+-------+-------------+-------------+-------------+-------------+-------------+---
--------+------------+-----------------+------------+------------+-----------------------+--------------
---+------------------+----------------+---------------------+--------------------+--------------+-----
------+--------------+
| %         | test    |       | Y           | Y           | Y           | Y           | Y           | Y
        | N          | Y               | Y          | Y          | Y                     | Y
    | Y                | Y              | N                   | N                  | Y            | Y
      | Y            |
| %         | test\_% |       | Y           | Y           | Y           | Y           | Y           | Y
        | N          | Y               | Y          | Y          | Y                     | Y
    | Y                | Y              | N                   | N                  | Y            | Y
      | Y            |
| 127.0.0.1 | db1     | user2 | Y           | Y           | Y           | N           | N           | N
        | N          | N               | N          | N          | N                     | N
    | N                | N              | N                   | N                  | N            | N
      | N            |
| %         | db1     | user3 | Y           | Y           | Y           | Y           | Y           | Y
        | N          | Y               | Y          | Y          | Y                     | Y
```

图 4-3-2　查看 db 表内的内容

步骤3： 查看db表内含有db字段的内容“select db from mysql.db;”，如图4-3-3所示。

```
mysql> select db from mysql.db;
+---------+
| db      |
+---------+
| db1     |
| test    |
| test\_% |
| db1     |
+---------+
4 rows in set (0.00 sec)
```

图 4-3-3　查看 db 表内含有 db 字段的内容

步骤4： 搜索查看多个字段“select db,user from mysql.db;”，如图4-3-4所示。

```
mysql> select db,user from mysql.db;
+----------+-------+
| db       | user  |
+----------+-------+
| db1      | user3 |
| test     |       |
| test\_%  |       |
| db1      | user2 |
+----------+-------+
4 rows in set (0.00 sec)
```

图 4-3-4　搜索查看多个字段

> **说明：**
> 搜索多个字段时，字段中间要用“,”隔开。

步骤5： 查询host为127.0的内容“select * from mysql.db where host like '127.0.%'\G;”，如图4-3-5所示。

```
mysql> select * from mysql.db where host like '127.0.%'\G;
*************************** 1. row ***************************
                 Host: 127.0.0.1
                   Db: db1
                 User: user2
          Select_priv: Y
          Insert_priv: Y
          Update_priv: Y
          Delete_priv: N
          Create_priv: N
            Drop_priv: N
           Grant_priv: N
      References_priv: N
           Index_priv: N
           Alter_priv: N
Create_tmp_table_priv: N
     Lock_tables_priv: N
     Create_view_priv: N
       Show_view_priv: N
  Create_routine_priv: N
   Alter_routine_priv: N
         Execute_priv: N
           Event_priv: N
         Trigger_priv: N
1 row in set (0.00 sec)
```

图 4-3-5　查询 host 为 127.0 的内容

步骤6： 向db1.t1中插入内容“insert into db1.t1 values (1, 'abc');”，如图4-3-6所示。

```
mysql> insert into db1.t1 values (1, 'abc');
Query OK, 1 row affected (0.04 sec)

mysql> select  * from db1.t1;
+------+------+
| id   | name |
+------+------+
|    1 | abc  |
|    1 | abc  |
+------+------+
2 rows in set (0.00 sec)
```

图 4-3-6　向 db1.t1 中插入内容

步骤7： 把id=1的字段内容更新成aaa“update db1.t1 set name='aaa' where id=1;”，

如图4-3-7所示。

```
mysql> update db1.t1 set name='aaa' where id=1;
Query OK, 2 rows affected (0.00 sec)
Rows matched: 2  Changed: 2  Warnings: 0

mysql> select * from db1.t1;
+----+------+
| id | name |
+----+------+
|  1 | aaa  |
|  1 | aaa  |
+----+------+
2 rows in set (0.00 sec)
```

图 4-3-7　把 id=1 的字段内容更新成 aaa

步骤8： 清空db1.t1表内的内容"truncate table db1.t1;"，如图4-3-8所示。

```
mysql> truncate table db1.t1;
Query OK, 0 rows affected (0.09 sec)

mysql> select * from db1.t1;
Empty set (0.00 sec)
```

图 4-3-8　清空 db1.t1 表内的内容

> **说明：**
> 清空后表的结构依然存在。

步骤9： 删除db1.t1表内的内容"drop table db1.t1;"，如图4-3-9所示。

```
mysql> drop table db1.t1;
Query OK, 0 rows affected (0.03 sec)

mysql>
mysql> desc db1.t1;
ERROR 1146 (42S02): Table 'db1.t1' doesn't exist
```

图 4-3-9　删除 db1.t1 表内的内容

> **说明：**
> 清空后连同表的结构一同删除。

步骤10： 清空db1.t1数据库"drop database db1;"，如图4-3-10所示。

```
mysql> drop database db1;
Query OK, 0 rows affected (0.05 sec)

mysql> show databases;
+--------------------+
| Database           |
+--------------------+
| information_schema |
| mysql              |
| performance_schema |
| test               |
+--------------------+
4 rows in set (0.00 sec)
```

图 4-3-10　删除 db1.t1 表内的内容

考评记录

姓名			完成日期	
序号	考核内容		标准分	评分
1	查看表内行数 "select count(*) from mysql.user;"		10	
2	查看 db 表内的内容 "select * from mysql.db;"		10	
3	查看 db 表内含有 db 字段的内容 "select db from mysql.db;"		10	
4	搜索查看多个字段 "select db,user from mysql.db;"		10	
5	查询 host 为 127.0 的内容 "select * from mysql.db where host like '127.0.%' \ G;"		10	
6	向 db1.t1 中插入内容 "insert into db1.t1 values (1, 'abc');"		10	
7	把 id=1 的字段内容更新成 aaa "update db1.t1 set name='aaa' where id=1;"		10	
8	清空 db1.t1 表内的内容 "truncate table db1.t1;"		10	
9	删除 db1.t1 表内的内容 "drop table db1.t1;"		10	
10	清空 db1.t1 数据库 "drop database db1;"		10	
	总评分		100	

任务实现心得：

任务实训

任务实训	新建一个数据库表，向里面插入 5 条数据，查询是否成功插入
任务目标	掌握 MySQL 的基本命令

任务 2　MySQL 数据库的备份与恢复

任务描述

情境描述	项目经理A对现有数据库中的数据量很满意，但是出于安全考虑，认为应该阶段性地对数据库备份，以防数据灾难的发生，所以组织项目组成员学习了数据库的备份和恢复知识
任务分解	分析上面的工作情境，将任务分解如下： （1）备份库； （2）恢复库； （3）备份表； （4）恢复表； （5）备份所有库； （6）只备份表结构
任务准备	初具规模的数据库，MySQL数据库的基本知识

视频

MySQL数据库的备份与恢复

任务目标

知识目标	掌握数据库的备份与恢复知识
技能目标	学习备份库、恢复库、备份表、恢复表、备份表结构
素质目标	耐心与细心：针对输入指令的报错问题进行细心的分析，MySQL数据库的相关指令能够耐心地完成

任务实现

步骤1： 备份库。

```
mysqldump -uroot -plinux mysql > /tmp/mysql.sql
```

步骤2： 恢复库。

```
mysql -uroot -plinux mysql < /tmp/mysql.sql
```

步骤3： 备份表。

```
mysqldump -uroot -plinux mysql user > /tmp/user.sql
```

步骤4： 恢复表。

```
mysql -uroot -plinux mysql < /tmp/user.sql
```

步骤1~步骤4的运行界面如图4-3-11所示。

图 4-3-11　备份库、表与恢复库、表

步骤5： 备份所有库。

```
mysqldump -uroot -plinux -A > /tmp/123.sql
```

步骤6： 只备份表结构。

```
mysqldump -uroot -plinux -d mysql > /tmp/mysql.sql
```

步骤5和步骤6的运行界面如图4-3-12所示。

图 4-3-12　备份所有库与备份表结构

学习笔记

任务考评

考评记录

姓名		完成日期	
序号	考核内容	标准分	评分
1	mysqldump -uroot -plinux mysql > /tmp/mysql.sql	10	
2	mysql -uroot -plinux mysql < /tmp/mysql.sql	20	
3	mysqldump -uroot -plinux mysql user > /tmp/user.sql	10	
4	mysql -uroot -plinux mysql < /tmp/user.sql	20	
5	mysqldump -uroot -plinux -A > /tmp/123.sql	20	
6	mysqldump -uroot -plinux -d mysql > /tmp/mysql.sql	20	
	总评分	100	

任务实现心得：

任务实训

任务实训	备份之前建立的 test 数据库中的 test1 表
任务目标	掌握备份的相关指令

单元 4　MySQL 主从配置

本单元包括四个任务：任务 1 是主配置，包括基础配置、编辑配置文件、重启 mysqld 服务、备份 MySQL 库、显示主机状态；任务 2 是从配置，包括基础配置、编辑配置文件、重启 mysqld 服务、在主上复制文件并检查是否一致、数据库配置；任务 3 是主从同步及相关配置参数；任务 4 是测试主从，包括在主上进入数据库以及在从上进入数据库。

学习目标

通过本单元的学习，使学生掌握 MySQL 主从配置的基本知识，培养学生自主进行 MySQL 主从配置的能力。

任务 1　主配置（安装完 MySQL 的虚拟机）

任务描述

情境描述	程序员A了解得知，大型网站为了缓解大量的并发访问，除了在网站实现分布式负载均衡之外，还会考虑如何减少数据库的连接。可以采用优秀的代码框架，进行代码的优化，采用优秀的数据缓存技术如memcached；如果资金丰厚，还可以架设服务器群，来分担主数据库的压力。本任务实现利用MySQL主从配置进行读写分离，减轻数据库压力
任务分解	分析以上工作情境，将任务分解如下： （1）基础配置； （2）编辑配置文件； （3）重启mysqld服务； （4）备份MySQL库； （5）显示主机状态
任务准备	（1）掌握基本MySQL知识； （2）安装完成MySQL的虚拟机

任务目标

知识目标	了解主从配置的必要性
技能目标	学会如何进行主从配置
素质目标	耐心与细心：针对输入指令的报错问题进行细心的分析，MySQL数据库的相关指令能够耐心地完成

任务实现

步骤1： 了解MySQL主从原理图（见图4-4-1）并进行基础配置。

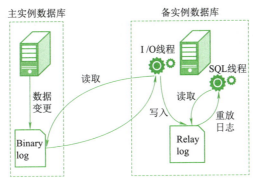

图 4-4-1 MySQL 主从原理

（1）主服务器上有一个log dump线程，用来和从的I/O线程传递binlog；

（2）从服务器上有两个线程，其中I/O线程用来同步主的binlog并生成relaylog，另外一个SQL线程用来把relaylog里面的SQL语句落地。

知识链接

MySQL 主从又称 Replication、AB 复制。简单讲就是 A 和 B 两台机器做主从后，在 A 上写数据，另外一台 B 也会跟着写数据，两者数据实时同步。

MySQL 主从是基于 binlog 的，主上须开启 binlog 才能进行主从。主从过程大致有三个步骤：

（1）主将更改操作记录到 binlog 中。
（2）从将主的 binlog 事件（SQL 语句）同步到从本机上并记录在 relaylog 的中继日志中。
（3）从根据 relaylog 中的 SQL 语句按顺序执行。

步骤2： 编辑配置文件，如图4-4-2所示。

```
# vim/etc/my.cnf
```

```
[root@localhost mysql]# vim /etc/my.cnf
# For advice on how to change settings please see
# http://dev.mysql.com/doc/refman/5.6/en/server-configuration-defaults.html
# *** DO NOT EDIT THIS FILE. It's a template which will be copied to the
# *** default location during install, and will be replaced if you
# *** upgrade to a newer version of MySQL.

[mysqld]
# Remove leading # and set to the amount of RAM for the most important data
# cache in MySQL. Start at 70% of total RAM for dedicated server, else 10%.
# innodb_buffer_pool_size = 128M

# Remove leading # to turn on a very important data integrity option: logging
# changes to the binary log between backups.
# log_bin

# These are commonly set, remove the # and set as required.
log_bin=linux01
basedir = /usr/local/mysql
datadir = /data/mysql
port = 3306
server_id = 12
socket =/tmp/mysql.sock
```

图 4-4-2 编辑配置文件

步骤3: 重启mysqld服务,如图4-4-3所示。

```
# /etc/init.d/mysqld restart
```

```
[root@localhost mysql]# /etc/init.d/mysqld restart
Shutting down MySQL.. SUCCESS!
Starting MySQL. SUCCESS!
```

图 4-4-3　重启 mysqld 服务

步骤4: 备份mysql库（加入环境变量），如图4-4-4所示。

```
# mysqldump -uroot mysql > /tmp/mysql.sql
```

```
[root@localhost ~]# mysqldump -uroot mysql > /tmp/mysql.sql
[root@localhost ~]# mysql -uroot -e "create database kei"
[root@localhost ~]# mysql -uroot kei < /tmp/mysql.sql
```

图 4-4-4　备份 mysql 库

步骤5: 创建一个库保存数据。

```
# mysql -uroot -e "create database kei"
```

将mysql库恢复成新建的库，作为测试数据。

```
# mysql -uroot kei < /tmp/mysql.sql
```

步骤6: 数据库配置。

```
# mysql -uroot
```

步骤7: 进入数据库（没有密码）。

```
> grant replication slave on *.* to 'repl' @192.168.37.13 identified by '123456';
```

步骤8: 创建用作同步数据的用户并赋予权限，如图4-4-5所示。

```
> flush tables with read lock;
```

```
mysql> grant replication slave on *.* to 'repl' @192.168.37.13 identified by 'password';
Query OK, 0 rows affected (0.00 sec)
```

图 4-4-5　创建用户并赋予权限

步骤9: 将表锁住,保持表内数据不变,如图4-4-6所示。

```
> show master status;
```

```
mysql> flush tables with read lock;
Query OK, 0 rows affected (0.00 sec)
```

图 4-4-6　锁住表

步骤10: 显示主机状态,如图4-4-7所示。

```
mysql> show master status;
+----------------+----------+--------------+------------------+-------------------+
| File           | Position | Binlog_Do_DB | Binlog_Ignore_DB | Executed_Gtid_Set |
+----------------+----------+--------------+------------------+-------------------+
| linux01.000004 |      120 |              |                  |                   |
+----------------+----------+--------------+------------------+-------------------+
1 row in set (0.00 sec)
```

图 4-4-7　显示主机状态

学习笔记

考评记录

姓名			完成日期	
序号	考核内容		标准分	评分
1	基础配置		20	
2	编辑配置文件		20	
3	重启 mysqld 服务		20	
4	备份 MySQL 库		20	
5	显示主机状态		20	
	总评分		100	

任务实现心得：

任务实训

任务实训	创建另外一个库来保存数据
任务目标	掌握数据库的主从配置

任务 2　从配置（安装完 MySQL 的虚拟机）

任务描述

情境描述	程序员A设计完主数据库之后还需要设置从数据库，才能完成主从配置
任务分解	分析上面的工作情境，将任务分解如下： （1）基础配置 （2）编辑配置文件 （3）重启mysqld服务 （4）在主上复制文件并检查是否一致 （5）数据库配置
任务准备	主数据库已经配置好，安装完MySQL的虚拟机

任务目标

知识目标	掌握从数据库的设置
技能目标	学会从数据库的配置
素质目标	耐心与细心：针对输入指令的报错问题进行细心的分析，MySQL数据库的相关指令能够耐心地完成

任务实现

步骤1： 基础配置。

```
# vi /etc/my.cnf
```

步骤2： 编辑配置文件，如图4-4-8所示。

```
# /etc/init.d/mysqld restart
```

```
[root@localhost ~]# vi /etc/my.cnf
# For advice on how to change settings please see
# http://dev.mysql.com/doc/refman/5.6/en/server-configuration-defaults.html
# *** DO NOT EDIT THIS FILE. It's a template which will be copied to the
# *** default location during install, and will be replaced if you
# *** upgrade to a newer version of MySQL.

[mysqld]
# Remove leading # and set to the amount of RAM for the most important data
# cache in MySQL. Start at 70% of total RAM for dedicated server, else 10%.
# innodb_buffer_pool_size = 128M

# Remove leading # to turn on a very important data integrity option: logging
# changes to the binary log between backups.
# log_bin

# These are commonly set, remove the # and set as required.
basedir = /usr/local/mysql
datadir = /data/mysql
port = 3306
server_id = 13
socket = /tmp/mysql.sock
```

4-4-8　编辑配置文件

步骤3： 重启mysqld服务，如图4-4-9所示。

```
# scp /tmp/mysql.sql root@192.168.37.13:/tmp/
```

```
[root@localhost mysql]# /etc/init.d/mysqld restart
Shutting down MySQL.. SUCCESS!
Starting MySQL. SUCCESS!
```

图 4-4-9　重启 mysqld 服务

步骤4： 在主上复制文件到从上，并在从上查看文件大小是否一致，如图4-4-10所示。

```
[root@localhost mysql]# scp /tmp/mysql.sql root@192.168.37.13:/tmp/
root@192.168.37.13's password:
mysql.sql                    100%  680KB  88.6MB/s   00:00
[root@localhost mysql]# ls -al /tmp/mysql.sql
-rw-r--r--. 1 root root 696710 Jul  6 21:15 /tmp/mysql.sql
```

图 4-4-10　在主上复制文件并在从上查看文件大小是否一致

创建一个和主一样的库，如图4-4-11所示。

```
# mysql -uroot -e "create database kei"
```

```
[root@localhost mysql]# mysql -uroot -e "create database kei"
-bash: mysql: command not found
[root@localhost mysql]# export PATH=$PATH:/usr/local/mysql/bin/
[root@localhost mysql]# mysql -uroot -e "create database kei"
```

图 4-4-11　创建一个和主一样的库

使用重定向方式，将mysql.sql数据库导入已创建的数据库kei中。将文件内容导入库。

```
# mysql -uroot kei < /tmp/mysql.sql
```

步骤5： 数据库配置。

```
# mysql -uroot
```

进入数据库（没有密码），如图4-4-12所示。

```
> change master to master_host='192.168.37.12',master_user='repl',master_password='123456',master_log_file='linux1.000001',master_log_pos=698861;
```

项目 4 MySQL 常用操作

```
[root@localhost mysql]# mysql -uroot
Welcome to the MySQL monitor.  Commands end with ; or \g.
Your MySQL connection id is 53
Server version: 5.6.45 MySQL Community Server (GPL)

Copyright (c) 2000, 2019, Oracle and/or its affiliates. All rights reserved.

Oracle is a registered trademark of Oracle Corporation and/or its
affiliates. Other names may be trademarks of their respective
owners.

Type 'help;' or '\h' for help. Type '\c' to clear the current input statement.

mysql> stop slave;
Query OK, 0 rows affected, 1 warning (0.00 sec)

mysql> change master to master_host='192.168.37.12',master_user='repl',master_password='password
',master_log_file='linux01.000004',master_log_pos=120;
Query OK, 0 rows affected, 2 warnings (0.00 sec)

mysql> start slave;
Query OK, 0 rows affected (0.00 sec)
```

图 4-4-12 进入数据库

在主上执行解锁表，如图4-4-13所示。

```
> unlock tables;
```

```
mysql> unlock tables;
Query OK, 0 rows affected (0.00 sec)
```

图 4-4-13 在主上执行解锁表

学习笔记

考评记录

姓名			完成日期	
序号	考核内容		标准分	评分
1	基础配置		25	
2	编辑配置文件		25	
3	重启mysqld服务		25	
4	在主上复制文件并检查是否一致显示主机状态		25	
	总评分		100	

任务实现心得：

任务实训

任务实训	修改主数据库中数据在从数据库中的存放位置
任务目标	熟悉从数据库的配置过程

任务 3　主从同步及相关配置参数

任务描述

情境描述	程序员A在配置好主从数据库之后还需要配置主从同步及相关的配置参数
任务分解	分析上面的工作情境，需要完成主从同步验证
任务准备	主从数据库已经配置完成

任务目标

知识目标	掌握主从同步及相关配置参数
技能目标	学会主从配置基本操作
素质目标	耐心与细心：针对输入指令的报错问题进行细心的分析，MySQL数据库的主从配置相关指令能够耐心地完成

任务实现

步骤： 主从同步验证。

从服务器上操作并执行命令（防火墙关闭），如图4-4-14所示。

```
> show slave status\G;
```

```
Slave_IO_Running: Yes
Slave_SQL_Running: Yes
Seconds_Behind_Master: 0
SL_Verify_Server_Cert: No
        Last_IO_Errno: 0
        Last_IO_Error:
       Last_SQL_Errno: 0
       Last_SQL_Error:
```

图 4-4-14　主从同步验证步骤 1

若出现Yes，如图4-4-14所示，即表示主从配置正常。
其中，主服务器主要配置参数如图4-4-15所示。

```
binlog-do-db=         //仅同步指定的库
binlog-ignore-db=     //忽略指定库
```

图 4-4-15　主从同步验证步骤 2

从服务器主要配置参数如图4-4-16所示。

```
replicate_do_db=
replicate_ignore_db=
replicate_do_table=
replicate_ignore_table=
replicate_wild_do_table=      //如test.%，支持涵配符%
replicate_wild_ignore_table=
```

图 4-4-16　主从同步验证步骤 3

学习笔记

考评记录

姓名		完成日期	
序号	考核内容	标准分	评分
1	主从同步验证	100	
	总评分	100	
任务实现心得：			

任务实训

任务实训	学习各个配置参数的含义
任务目标	深刻了解主从配置过程

任务 4　测试主从

任务描述

情境描述	测试工程师B需要测试主从数据库的连接情况，在网上查阅了相关资料，知道了如何测试主从数据库
任务分解	分析上面的工作情境，将任务分解如下： （1）在主上进入数据库； （2）在从上进入数据库
任务准备	配置好主从服务器

任务目标

知识目标	了解测试主从数据库的方法
技能目标	学会主服务器与从服务器的测试方法
素质目标	耐心与细心：针对输入指令的报错问题进行细心的分析，MySQL数据库的相关指令能够耐心地完成

任务实现

步骤1： 主服务器上。

```
# mysql -uroot -p密码
```

在主上进入数据库，如图4-4-17所示。

```
> select count(*) from db;
> truncate table db;
```

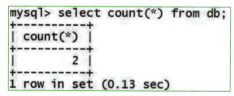

图 4-4-17　在主上进入数据库

步骤2： 从服务器上。

```
# mysql -uroot kei
```

在从上进入数据库

> select count(*) from db;

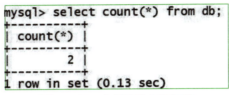

图 4-4-18　在从上进入数据库

学习笔记

考评记录

姓名			完成日期	
序号	考核内容		标准分	评分
1	在主上进入数据库		50	
2	在从上进入数据库		50	
	总评分		100	

任务实现心得：

任务实训

任务实训	测试 test 数据库的主从情况
任务目标	掌握测试的相关知识

项目 5
网站优化与服务器优化

本项目主要介绍了网站优化与服务器优化。通过本项目的学习可以了解如何优化网站和服务器。共分为两个单元来介绍，分别介绍了网站和服务器的基础知识、网站速度的检测方法、注册和使用 DNSPOD 与接入 CDN 厂商。

云计算应用运维实战

单元 1　了解基础知识与检测网站打开速度

本单元包括两个任务：任务 1 是了解网站优化与服务器优化的基础知识，包括了解服务器、VPS 与空间、了解网站结构的演变；任务 2 是检测网站打开速度，包括下载谷歌（Google）浏览器与搜索 360 网站测速，打开该网址，输入 IP。

学习目标

通过本单元的学习，使学生掌握检测网站打开速度和网站优化与服务器优化的基础知识，培养学生进行检测网站打开速度的能力。

任务 1　了解网站优化与服务器优化的基础知识

任务描述

视频
常用网站优化的方法

情境描述	项目经理 A 是某网站的项目负责人，他发现网站还有进行优化的空间，所以对网站进行了优化。 常用网站优化方法有： （1）架构的调整。比如，只有三台 Web，那访问会很快达到上限，网站访问速度就会减慢。如果此时加上几台 Web 服务器，使用负载均衡来分发流量，速度自然又会提高。当然，也可以调整缓存服务器，这都是视具体情况而定的。 （2）硬件的优化。增加服务器内存、CPU。 （3）应用程序的优化。网站使用的是 LAMP，其中 L 是 Linux，A 是 Apache，M 是 MySQL，P 是 PHP，因此可以针对它们进行优化。Linux 不同版本有不同的内核，可以利用版本进行内核优化，或是对 Apache、MySQL、PHP 三者进行优化。 （4）程序的优化。可以通过监控应用程序文件查看哪些文件执行得慢，然后找出写得不好的地方并修改。 进行网站优化与服务器优化的前提是要对网站优化与服务器优化的基础知识进行了解，下面介绍服务器、VPS 与空间与网站结构演变
任务分解	分析上面的工作情境，将任务分解如下： （1）了解服务器、VPS 与空间； （2）了解网站结构的演变
任务准备	拥有一个乐于学习的心态

任务目标

知识目标	了解网站优化与服务器优化的基础知识
技能目标	了解服务器、VPS 与空间、网站结构的演变
素质目标	细心与耐心：认真、仔细、耐心学习相关知识

5-2

任务实现

步骤1： 了解服务器、VPS与空间。

视频
服务器、VPS、空间的介绍

（1）服务器。服务器就是一台计算机，配备了一系列提供Web服务所必需的软件。服务器有自己专用的CPU主板、内存等，很多都是为了保证稳定性的。

（2）VPS。VPS称为虚拟个人服务器，又称虚拟独立主机。它和虚拟主机的共同点是采用软件技术把一台物理服务器分成多个具备服务器功能的账户，但是它比虚拟主机更高级，账户之间更独立，权限更多。

（3）空间。网站空间又称网页空间、建站空间，是指托管网站的服务器，包括服务器、虚拟主机、VPS。网络空间就是指网页存放的那个地方，物理中对应的就是服务器上的硬盘。

步骤2： 了解网站结构的演变。

视频
网站结构的演变过程

通过DNS访问IP来访问网站，经常由于访问量过大，网站承受不住。此时可以分离Web端和数据库端，访问量再增加时就要增加Web，然后使用DNS连接到负载均衡服务器，使流量平均到每个Web端。如果此时负载均衡服务器故障，底下的Web服务器就访问不到了，此时还得增加负载均衡服务器，第一台故障了，会自动访问第二台，实现负载均衡高可用。

如果要继续降低Web服务器的负荷，还可以在Web服务器与负载均衡服务器之间加一个缓存层，缓存层用于保存静态资源。如果用户访问网页而缓存层有这个静态资源，那么就不需要到Web服务器上去读取了；如果缓存层没有这个静态资源，那么当用户读取完这个资源后，缓存层就会保存这个资源。

例如，如果网站是一个论坛，那访问帖子时就要访问数据库服务器了。如果数据库服务器故障了，帖子就看不了了。那怎么优化？可以做一个MySQL AB复制，即同步备份，如果第一台数据库做出了更改，那第二台服务器也会做出相应的修改。也可以在Web服务器与数据库服务器之间加上Memory Cache（内存缓存）服务器，这个服务器用于存储一些比较热门的帖子，用于降低数据库服务器的压力。

总结：网站演变，由一台服务器变成多台服务器。

网站结构的演变分层如图5-1-1所示。

图 5-1-1 网站结构的演变分层

考评记录

姓名			完成日期	
序号	考核内容		标准分	评分
1	服务器、VPS 与空间基础知识		50	
2	网站结构的演变		50	
	总评分		100	

任务实现心得：

任务实训

任务实训	根据所学内容，举例说明网站的优化方案
任务目标	掌握网站优化策略

任务 2　检测网站打开速度

任务描述

情境描述	程序员A认为网页载入速度对于一个网站来讲很关键，而且Google已经将一个网站的载入速度列入了网站关键字排名的考虑因素当中，也就是说如果网站有足够的内容，而且载入速度比别的网站更快一步，那么就可以获得更好的排名。他准备测试一下自己的网站，然后提高网站访问速度
任务分解	分析上面的工作情境，将任务分解如下： （1）下载谷歌（Google）浏览器； （2）搜索360网站测速，打开该网址，输入IP
任务准备	（1）掌握如何下载计算机软件； （2）掌握计算机搜索技能

任务目标

知识目标	知道如何检测网站打开速度
技能目标	能独立检测网站打开速度
素质目标	细心与耐心：具备独立的检测和分析能力

任务实现

步骤1： 下载谷歌（Google）浏览器，按【F12】键，之后打开某个网站，会显示网站加载时间，如图5-1-2所示。

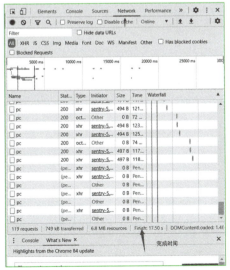

图 5-1-2　检测网站打开速度步骤 1

步骤2： 测试各地打开某个网站的速度。例如，搜索360网站测速（见图5-1-3），打开该网址，输入IP。

图 5-1-3　检测网站打开速度步骤2

学习笔记

考评记录

姓名			完成日期	
序号	考核内容		标准分	评分
1	下载谷歌（Google）浏览器		50	
2	搜索 360 网站测速，打开该网址，输入 IP		50	
	总评分		100	

任务实现心得：

任务实训

任务实训	测试搜狐网站的速度
任务目标	了解网站如何测速

单元 2　注册和使用 DNSPOD 与接入 CDN 厂商

DNS原理解析

注册和使用 DNSPOD

本单元包括两个任务：任务 1 是注册和使用 DNSPOD，包括百度搜索 DNSPOD 并注册、添加域名、将 DNS 解析服务器设置修改为对应记录值、添加监控、查询域名相关注册商；任务 2 是接入 CDN 厂商，包括创建 CDN 与将显示的 CNAME 配置在 DNSPOD 中，类型为 CNAME。

学习目标

通过本单元的学习，使学生掌握注册和使用 DNSPOD 与接入 CDN 厂商的基础知识，培养学生自主进行注册和使用 DNSPOD 与接入 CDN 厂商的能力。

任务 1　注册和使用 DNSPOD

任务描述

情境描述	程序员 A 在给公司新员工培训的过程中发现，很多新员工还不太了解 DNS 的原理与使用，所以他对新员工介绍了以下内容。 　　通常访问一个网站，不管它是服务器、VPS 还是空间，都要提供一个对外的公网 IP，访问这个网站可以通过这个 IP 来实现。但是，IP 地址并不好记，这时就可以通过域名来访问。域名一般比较容易记忆。那这个工作是怎么完成的？这就需要一个 DNS 服务器了。访问域名时，这个 DNS 服务器会自动识别域名，将其接到对应的 IP 地址上去。 　　域名是一个树状结构，最开头是根域，根域下又分为 .com、.cn、.net、.org 等域，这些统称一级域。而在一级域中配置的 NS 服务器指向各个网站，如平常访问的 www.baidu.com 中域名为 baidu.com.（最后的点可以不写），它属于 .com 一级域。www 称为主机头。由主机头和域名构成的地址才能成为一个完整的网址。也有人将 www.baidu.com 等域名称为二级域名。现在在国内用得比较多的 NS 服务器是 DNSPOD，它是一个第三方运营商，现在是腾讯的一个子公司，可以给用户提供一台速度快的 NS 服务器。当访问域名的时候，是由 NS 服务器来帮助将域名和 IP 进行连接的。比如，用户访问 www.baidu.com 时，它会从根域开始往下找，找到对应的域名服务器，再帮用户解析到相应 IP
任务分解	分析上面的工作情境，将任务分解如下： （1）百度搜索 DNSPOD 并注册； （2）添加域名； （3）将 DNS 解析服务器修改为对应记录值； （4）添加监控； （5）可以查询域名相关注册商
任务准备	（1）掌握计算机搜索技能； （2）拥有一台计算机

项目 5　网站优化与服务器优化

任务目标

知识目标	掌握查找域名注册商的方法
技能目标	能独立注册和使用DNSPOD
素质目标	细心与耐心：能认真学习相关知识

任务实现

步骤1： 百度搜索DNSPOD，可以看到它的一级域名是.cn，这是中国的域名，先打开它（见图5-2-1），然后进行注册。

然后，打开管理控制台。新注册的用户实名认证即可。

图 5-2-1　注册和使用 DNSPOD 步骤 1

步骤2： 添加域名，如图5-2-2所示。

图 5-2-2　注册和使用 DNSPOD 步骤 2

步骤3： 添加完发现状态是错误的，因此需要去相应的域名注册商，将DNS解析服务器修改为对应记录值，如图5-2-3所示。

图 5-2-3　注册和使用 DNSPOD 步骤 3

5-9

步骤4： 修改完之后，添加监控，如图5-2-4所示。

图 5-2-4　注册和使用 DNSPOD 步骤 4

步骤5： 如果不知道域名相应的注册商，可以百度whois（见图5-2-5）。可以查询域名相关注册商，百度搜索注册商就可以知道它是哪一家的，如图5-2-6所示。

图 5-2-5　注册和使用 DNSPOD 步骤 5

图 5-2-6　注册和使用 DNSPOD 步骤 6

 任务考评

考评记录

姓名		完成日期	
序号	考核内容	标准分	评分
1	百度搜索 DNSPOD 并注册	20	
2	添加域名	20	
3	将 DNS 解析服务器设置修改为对应记录值	20	
4	添加监控	20	
5	可以查询域名相关注册商	20	
	总评分	100	

任务实现心得：

任务实训

任务实训	重新注册一个 DNSPOD，完成整个过程
任务目标	了解 DNSPOD 的使用方法

任务 2　接入 CDN 厂商

任务描述

视频
CDN原理解析

情境描述	程序员B今天需要给新员工培训CDN的相关知识，所以他前期做了如下的知识准备。 CDN是内容分发网络，缓存源端数据，供受地域限制的人访问。 例如，如果有一台Web端，也就是用户访问的源端在本地，那么访问的速度就会比较快；如果没有CDN，其他国家和地区的人访问此地址，速度自然就会比较慢。而这种问题可以通过CDN来缓解。 CDN架构是由一个智能DNS加上若干个节点组成。访问时，由这个智能DNS来判断以用户的网络访问哪一个节点的速度最快，它就会把用户解析到那一个节点去。 而如果没有CDN，那么用户访问网站时，都要到源端去获取资源，这就会大大地增加源站负担。而拥有了CDN可以减少对源端的负担，同时也减少了源端遭受攻击的可能性，因为它访问的是节点IP，而不是源端IP。 程序员A他准备接入CDN厂商
任务分解	分析上面的工作情境，将任务分解如下： （1）创建CDN； （2）将显示的CNAME配置在DNSPOD中，类型为CNAME
任务准备	（1）拥有一个自己的网站； （2）掌握网站基本知识

任务目标

知识目标	掌握如何接入CDN
技能目标	能独立接入CDN
素质目标	细心与耐心

任务实现

步骤1： 创建CDN。

（1）单击控制台，如图5-2-7所示。

视频
接入CDN厂商

图 5-2-7　创建 CDN 步骤 1

(2) 单击"云产品→SDN"菜单命令，如图5-2-8所示。

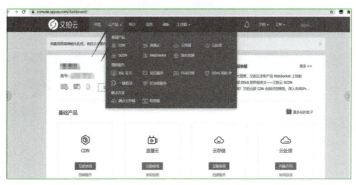

图 5-2-8　创建 CDN 步骤 2

(3) 单击"创建服务"按钮，如图5-2-9所示。

图 5-2-9　创建 CDN 步骤 3

•服务名称：唯一标识服务，如image-upyun-com，一个服务下面可以绑定多个自有域名。

•加速域名：填写此次需要配置的加速域名。

•应用场景：可选项包括网页图片、文件下载、音视频点播、动态内容、全站加速。

图 5-2-10　创建 CDN 步骤 4

- 回源协议：对回源协议进行配置，默认为HTTP，配置选项包括HTTP、HTTPS、协议跟随。
- 线路配置：此处包括回源地址、回源端口号的设置。

图 5-2-11　创建 CDN 步骤 5

（4）服务创建成功后，操作界面会提示CDN加速服务创建成功，并会自动跳转到该服务的"功能配置"界面，如图5-2-12所示。

图 5-2-12　创建 CDN 步骤 6

步骤2： 将显示的CNAME配置在DNSPOD中，类型为CNAME，如图5-2-13所示。

图 5-2-13　创建 CDN 步骤 7

申请UPYUN联盟和阿里云的注意事项：

可以加入又拍云联盟来获取一些免费资源，如图5-2-14和图5-2-15所示。

图 5-2-14　创建 CDN 步骤 8

图 5-2-15　创建 CDN 步骤 9

最后如果使用的是阿里云的服务器，在使用CDN代理的时候要去阿里云将代理节点的IP加入云盾的防火墙白名单内。只有阿里云才需要这步操作，如果是腾讯云的话则不需要这一步操作。

考评记录

姓名		完成日期	
序号	考核内容	标准分	评分
1	创建 CDN	70	
2	将显示的 CNAME 配置在 DNSPOD 中,类型为 CNAME	30	
	总评分	100	

任务实现心得:

任务实训	独立将自己的网站接入 CDN
任务目标	了解 CDN 的接入原理

项目 6
Tomcat 搭建配置

本项目主要介绍了 Tomcat 的相关知识。Tomcat 是开发人员常用的部署服务器。本项目共分为两个单元介绍,分别介绍了 JDK 与 Tomcat 的安装、配置 Tomcat 来进行监听等相关操作。

视频

Tomcat介绍

单元 1　安装 JDK 与安装 Tomcat

本单元包括两个任务：任务 1 是安装 JDK，包括下载 JDK、安装 JDK、修改环境变量、检查 JDK 安装情况；任务 2 是安装 Tomcat，包括环境准备、修改主机名、安装 Tomcat、启动 Tomcat5 测试。

视　频

安装JDK

学习目标

通过本单元的学习，使学生掌握安装 JDK 与安装 Tomcat 的基础知识，培养学生自主安装 JDK 与安装 Tomcat 的能力。

任务 1　安装 JDK

任务描述

情境描述	公司近期新接了一个以 Java 作为后台开发语言的网站开发，而程序员需要进行 JDK 的安装。使用 Java 程序编写的网站是用 Tomcat+JDK 来运行的。Tomcat 是一个中间件，真正起作用的、解析 Java 脚本的是 JDK。JDK（Java Development Kit）是整个 Java 的核心，它包含了 Java 运行环境和 Java 相关工具以及 Java 基础库。最主流的 JDK 为 Sun 公司发布的 JDK，除此之外，IBM 公司也发布有 JDK，CentOS 上也可以用 yum 安装 OpenJDK。要先安装 Java，才能安装 Tomcat。安装 Java 之前需要安装 JDK，然后进行环境变量控制。
任务分解	分析上面的工作情境，将任务分解如下： （1）下载 JDK； （2）安装 JDK； （3）修改环境变量； （4）检查 JDK 安装情况
任务准备	（1）掌握 Java 基本知识； （2）拥有安装 JDK 的硬件资源配置

任务目标

知识目标	了解安装 JDK 的必要性
技能目标	学会如何安装 JDK
素质目标	耐心与细心：针对输入指令的报错问题进行细心的分析，安装 JDK 的相关指令能够耐心地完成

任务实现

步骤1： 下载JDK。

官网下载地址：

http://www.oracle. com/echnetwork/java/javase/downloads/jdk8-downloads-2133151.html

这个下载地址不能在Linux虚拟机里使用wget命令下载，只能先通过浏览器下载到本机上，然后再上传到Linux。本次下载的版本为jdk1.8，将虚拟机连接CRT，通过CRT上传JDK到/usr/local/src/目录下。

步骤2： JDK安装。

部署Tomcat环境需要JDK软件环境。进入/usr/local/src/目录下，解压刚刚上传的jdk-8u211-linux-x64.tar.gz，将解压后的文件移动到/usr/local/目录下，重命名为jdk1.8。然后使用ls命令查看/usr/local/目录下是否有jdk1.8目录。

命令如下：

```
[root@tomcat ~]#cd  /usr/local/src/
[root@tomcat src]#tar zxf jdk-8u211-linux-x64.tar.gz
[root@tomcat src]#mvjdk1.8.0_211/   /usr/local/jdk1.8
[root@tomcat src]#ls  /usr/local/
bin  etc  games  include  jdk1.8  lib  lib64  libexecsbin  share  src
[root@tomcat src]#
```

步骤3： 修改环境变量。

设置环境变量，编辑/etc/profile文件，并使其立即生效。

```
[root@tomcat src]#vim  /etc/profile
//将以下内容添加到文件
……
JAVA_HOME=/usr/local/jdk1.8/
JAVA_BIN=/usr/local/jdk1.8/bin
JRE_HOME=/usr/local/jdk1.8/jre
PATH=$PATH:/usr/local/jdk1.8/bin:/usr/local/jdk1.8/jre/bin
CLASSPATH=/usr/local/jdk1.8/jre/lib:/usr/local/jdk1.8/lib:/usr/local/jdk1.8/jre/lib/charsets.jar
    [root@tomcat src]#source/etc/profile
```

步骤4： 检查JDK安装情况。

配置完成环境变量后，使用java -version命令检查是否安装成功，如果显示结果带有java version "1.8.0_211" 字样，和jdk-8u211-linux-x64.tar.gz包的版本相对应，则证明安装成功。命令如下：

```
[root@tomcat src]# java -version
java version "1.8.0_211"
Java(TM) SE Runtime Environment (build 1.8.0_211-b12)
Java HotSpot(TM) 64-Bit Server VM (build 25.211-b12, mixed mode)
[root@tomcat src]#
```

这里有可能出现的不是上面的这种,而是如下所示。

```
[root@tomcat src]# java -version
openjdk version "1.8.0_242"
OpenJDK Runtime Environment (build 1.8.0_242-b08)
OpenJDK 64-Bit Server VM (build 25.242-b08, mixed mode)
```

如果发现不是自己安装的JDK,这是因为系统自带OpenJDK或者以前安装过OpenJDK。使用witch命令,查看现在Java的所在目录:

```
[root@tomcat src]# which java
/usr/bin/java
```

如果结果为/usr/bin/java,则说明这是系统自带的OpenJDK。这时,为了实验的一致性,把原来的Java目录重命名为java_bak,并使用source命令再次使环境变量生效,然后使用java -verison命令查看是否有java version "1.8.0_211" 字样。命令如下:

```
[root@tomcat src]# mv /usr/bin/java /usr/bin/java_bak
[root@tomcat src]# source /etc/profile
[root@tomcat src]#java -version
java version "1.8.0_211"
Java(TM) SE Runtime Environment (build 1.8.0_211-b12)
Java HotSpot(TM) 64-Bit Server VM (build 25.211-b12, mixed mode)
[root@tomcat src]#
```

考评记录

姓名		完成日期	
序号	考核内容	标准分	评分
1	下载 JDK	25	
2	安装 JDK	25	
3	修改环境变量	25	
4	检查 JDK 安装情况	25	
	总评分	100	

任务实现心得：

任务实训

任务实训	在自己的计算机安装 JDK
任务目标	掌握安装 JDK 的必要性与原理

任务 2　安装 Tomcat

任务描述

视频 安装Tomcat

情境描述

程序员A接下来的工作是部署服务器，他选用的是Tomcat。Tomcat是Apache软件基金会（Apache Software Foundation）Jakarta项目中的一个核心项目，由Apache、Sun和其他一些公司及个人共同开发而成。它受到了Java爱好者的喜爱，并得到了部分软件开发商的认可，成为目前比较流行的Web应用服务器。Tomcat服务器是一个免费的开放源代码的Web 应用服务器，属于轻量级应用服务器，在中小型系统和并发访问用户不是很多的场合下被普遍使用，是开发和调试JSP程序的首选。目前有很多网站是用Java编写的，所以解析Java程序就必须由相关的软件完成，Tomcat就是其中之一。由于 Tomcat 本身内含了一个 HTTP 服务器，它也可以被视作一个单独的 Web 服务器。但是，不能将 Tomcat 和 Apache HTTP 服务器混淆，Apache HTTP 服务器是一个用 C 语言实现的 HTTP Web 服务器；这两个 HTTP Web Server 不是捆绑在一起的。Tomcat 包含了一个配置管理工具，也可以通过编辑XML格式的配置文件来进行配置

任务分解

分析上面的工作情境，将任务分解如下：
(1) 环境准备；
(2) 修改主机名；
(3) 安装Tomcat；
(4) 启动Tomcat；
(5) 测试

任务准备

掌握Tomcat基础知识

任务目标

知识目标	掌握安装Tomcat的必要性
技能目标	学会如何安装Tomcat
素质目标	耐心与细心：针对输入指令的报错问题进行细心的分析，安装Tomcat的相关指令能够耐心地完成

任务实现

步骤1: 环境准备。

规划节点，本次实验为单节点部署：

节点IP	主机名	节点
192.168.174.155	tomcat	Tomcat

使用VMWare Workstation软件安装CentOS 7.2操作系统，本案例镜像使用CentOS-7-

x86_64-DVD-1511.iso。

关闭防火墙并设置开机不自启，配置SElinux规则。

```
[root@localhost ~]# systemctl stop firewalld.service
[root@localhost ~]# systemctl disable firewalld.service
Removed symlink
/etc/systemd/system/dbus-org.fedoraproject.FirewallD1.service.
Removed symlink /etc/systemd/system/basic.target.wants/firewalld.service.
[root@localhost ~]# setenforce 0
```

步骤2： 使用hostnamectl命令修改主机名。命令如下：

```
[root@localhost ~]# hostnamectl set-hostname tomcat
// 修改完后，按【Ctrl+D】组合键退出后重新连接
[root@tomcat ~]#
```

步骤3： 安装Tomcat。

去官网下载合适的Tomcat版本（本案例已经下载好了），使用CRT上传到虚拟机/usr/local/src/目录下，然后解压，将解压后的文件移动到/usr/local目录下，并命名为tomcat。命令如下：

```
[root@tomcat ~]# cd /usr/local/src/
[root@tomcat src]# tar zxf apache-tomcat-9.0.21.tar.gz
[root@tomcat src]# mv apache-tomcat-9.0.21  /usr/local/tomcat
[root@tomcat src]#
```

步骤4： 启动Tomcat。

本次下载解压的包是二进制包，不用去编译，使用/usr/local/tomcat/bin/startup.sh命令启动Tomcat。命令如下：

```
[root@tomcat src]# /usr/local/tomcat/bin/startup.sh
Using CATALINA_BASE:   /usr/local/tomcat
Using CATALINA_HOME:   /usr/local/tomcat
Using CATALINA_TMPDIR: /usr/local/tomcat/temp
Using JRE_HOME:        /usr/local/jdk1.8
Using CLASSPATH:
/usr/local/tomcat/bin/bootstrap.jar:/usr/local/tomcat/bin/tomcat-juli.jar
Tomcat started.
[root@tomcat src]#
```

步骤5： 测试。

用netstat命令监听Java相关服务端口，查看是否有以下端口存在（8009、8080、

8005),如果存在则证明Tomcat服务启动成功。命令如下:

```
[root@tomcat src]# netstat -lnpt | grep java
tcp6       0      0 :::8009              :::*         LISTEN      49228/java
tcp6       0      0 :::8080              :::*         LISTEN      49228/java
tcp6       0      0 127.0.0.1:8005       :::*         LISTEN      49228/java
[root@tomcat src]#
```

三个端口8009、8005和8080的意义如下:

- 8080为提供Web服务的端口;
- 8005为管理端口;
- 8009端口为第三方服务调用的端口,比如httpd和Tomcat结合时会用到。

打开浏览器,在地址栏中输入http://IP:8080/(这里的IP为虚拟机的IP地址,此处IP地址为192.168.174.155),可以看到Tomcat的默认页面,如图6-1-1所示。

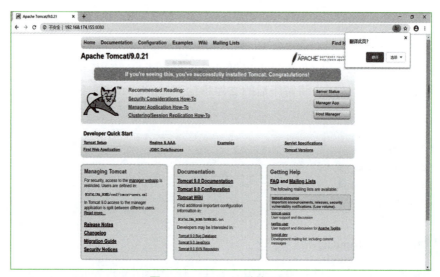

图 6-1-1　Tomcat 默认页面

学习笔记

项目 6　Tomcat 搭建配置

 任务考评

考评记录

姓名		完成日期	
序号	考核内容	标准分	评分
1	安装 Tomcat	40	
2	启动 Tomcat	30	
3	测试	30	
	总评分	100	

任务实现心得：

任务实训

任务实训	成功完成安装 Tomcat
任务目标	掌握 Tomcat 的安装原理

6-9

云计算应用运维实战

单元 2　配置Tomcat监听80端口、虚拟主机和生成日志

本单元包括三个任务：任务1是配置Tomcat监听80端口，包括修改配置文件、测试、浏览器访问；任务2是配置Tomcat虚拟主机，包括查看默认虚拟主机、新增虚拟主机、appBase部署Java应用、docBase部署Java应用；任务3是配置Tomcat生成日志，包括配置生成日志与查看日志。

学习目标

通过本单元的学习，使学生掌握配置Tomcat监听80端口、虚拟主机以及日志的基础知识，培养学生自主配置Tomcat监听80端口、虚拟主机以及配置Tomcat生成日志的能力。

任务 1　配置 Tomcat 监听 80 端口

任务描述

视频
配置Tomcat
监听80端口

情境描述	程序员A了解得知Tomcat默认的访问监听端口是8080，但是在网址栏输入IP再加上端口难免有些麻烦；同时，如果使用IP进行解析，只能解析到对应的IP，无法直接通过浏览器默认的80端口进行访问，因而可以修改Tomcat默认监听的端口为80，这样就可以方便地通过浏览器进行访问了。本任务总结如何配置Tomcat监听80端口
任务分解	分析上面的工作情境，将任务分解如下： （1）修改配置文件； （2）测试； （3）浏览器访问
任务准备	完成Tomcat基础环境安装

任务目标

知识目标	了解配置Tomcat监听80端口的必要性
技能目标	学会如何配置Tomcat监听80端口
素质目标	耐心与细心：针对输入指令的报错问题进行细心的分析，配置Tomcat监听80端口的相关指令能够耐心地完成

任务实现

▶ 步骤1： 修改配置文件。

Tomcat 监听的是8080端口，如果想要直接访问，不加8080端口（默认输入一个IP

或者域名，访问80端口时是可以省略IP后端口号的），就需要配置Tomcat监听80端口，Tomcat是支持端口自定义的。

编辑Tomcat配置文件server.xml。命令如下：

```
[root@tomcat ~]# vim /usr/local/tomcat/conf/server.xml
//直接搜索8080找到如下内容
...
    <Connector port="8080" protocol="HTTP/1.1"
               connectionTimeout="20000"
               redirectPort="8443" />
//将这里的8080直接改成80
...
[root@tomcat ~]#
```

修改完配置文件，接下来就是重启Tomcat服务，Tomcat服务是不支持以restart方式重启服务的，所以要想重启服务必须先关闭服务，使用/usr/local/tomcat/bin/shutdown.sh命令，然后再启动服务，使用/usr/local/tomcat/bin/startup.sh命令。

```
[root@tomcat ~]# /usr/local/tomcat/bin/shutdown.sh
Using CATALINA_BASE:   /usr/local/tomcat
Using CATALINA_HOME:   /usr/local/tomcat
Using CATALINA_TMPDIR: /usr/local/tomcat/temp
Using JRE_HOME:        /usr/local/jdk1.8
Using CLASSPATH:       /usr/local/tomcat/bin/bootstrap.jar:/usr/local/tomcat/bin/tomcat-juli.jar
[root@tomcat ~]#
[root@tomcat ~]# /usr/local/tomcat/bin/startup.sh
Using CATALINA_BASE:   /usr/local/tomcat
Using CATALINA_HOME:   /usr/local/tomcat
Using CATALINA_TMPDIR: /usr/local/tomcat/temp
Using JRE_HOME:        /usr/local/jdk1.8
Using CLASSPATH:       /usr/local/tomcat/bin/bootstrap.jar:/usr/local/tomcat/bin/tomcat-juli.jar
Tomcat started.
[root@tomcat ~]#
```

知识链接

网站绑定域名后直接通过域名访问使用的是 80 端口，因此 Tomcat 须监听 80 端口，而为了安全起见 Tomcat 一般不用 root 身份运行，综上，需要以普通用户来运行监听 80 端口的 Tomcat。此时就会启动失败，报没有权限，因为只有 root 身份才能监听 1024 以下的熟知端口。

步骤2: 测试。

重启服务之后,使用netstat命令监听Java相关服务端口来查看是否启动成功。命令如下:

```
[root@tomcat ~]# netstat -plnt | grep java
tcp6    0    0 :::8009            :::*    LISTEN    47709/java
tcp6    0    0 127.0.0.1:8005     :::*    LISTEN    47709/java
[root@tomcat ~]#
```

如果启动成功,则会出现三个端口(8009、8005、80)。

可以看到,这里并没有80端口出现,这是因为这台机器之前安装过LNMP。Nginx的默认端口也是80,这就会和刚刚设置的Tomcat端口发生冲突。这时,需要关闭Nginx服务,重新启动Tomcat服务。

```
[root@tomcat ~]# service nginx stop
Stopping nginx (via systemctl):  [  OK  ]
//出现OK字样,表示Nginx服务关闭成功
[root@tomcat ~]# /usr/local/tomcat/bin/shutdown.sh
Using CATALINA_BASE:   /usr/local/tomcat
Using CATALINA_HOME:   /usr/local/tomcat
Using CATALINA_TMPDIR: /usr/local/tomcat/temp
Using JRE_HOME:        /usr/local/jdk1.8
Using CLASSPATH:       /usr/local/tomcat/bin/bootstrap.jar:/usr/local/tomcat/bin/tomcat-juli.jar
[root@tomcat ~]# /usr/local/tomcat/bin/startup.sh
Using CATALINA_BASE:   /usr/local/tomcat
Using CATALINA_HOME:   /usr/local/tomcat
Using CATALINA_TMPDIR: /usr/local/tomcat/temp
Using JRE_HOME:        /usr/local/jdk1.8
Using CLASSPATH:       /usr/local/tomcat/bin/bootstrap.jar:/usr/local/tomcat/bin/tomcat-juli.jar
Tomcat started.
```

这时,再次使用netstat命令监听Java相关服务端口,查看三个端口是否存在。命令结果如下:

```
[root@tomcat ~]# netstat -plnt | grep java
tcp6    0    0 :::8009            :::*    LISTEN    47873/java
tcp6    0    0 :::80              :::*    LISTEN    47873/java
tcp6    0    0 127.0.0.1:8005     :::*    LISTEN    47873/java
[root@tomcat ~]#
```

可以看到这次三个服务端口都存在,证明Tomcat服务启动成功。

步骤3： 浏览器访问。

配置文件修改完成，且Tomcat服务启动成功后，打开浏览器在地址栏中输入http://192.168.174.155/（注意IP），这时又会看见Tomcat默认页面，如图6-2-1所示。

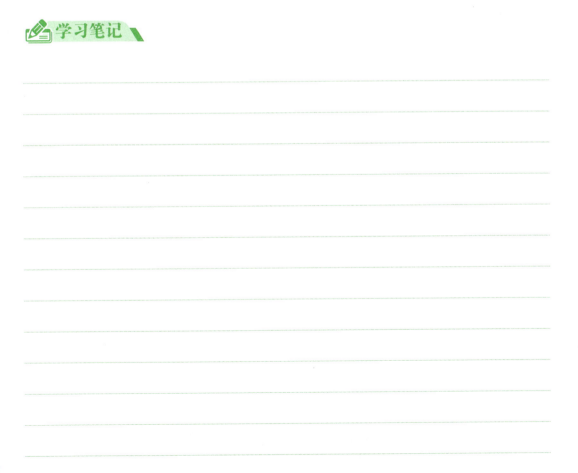

图 6-2-1 　 Tomcat 默认页面

学习笔记

任务考评

考评记录

姓名			完成日期	
序号	考核内容		标准分	评分
1	修改配置文件		40	
2	测试		30	
3	浏览器访问		30	
	总评分		100	

任务实现心得：

任务实训

任务实训	自行独立配置 Tomcat 监听 80 端口
任务目标	掌握配置 Tomcat 监听 80 端口的方法

任务 2　配置 Tomcat 虚拟主机

📋 任务描述

情境描述	为了响应公司开发需要，程序员A需要配置虚拟主机。Tomcat服务器允许用户在同一台计算机上配置多个Web站点，在这种情况下，需要为每个Web站点，配置不同的主机名即配置虚拟主机。 　　现实生活中，为了提高硬件资源的利用率，有很多网站通过配置虚拟主机的方式实现服务器的共享。本任务总结如何配置Tomcat虚拟主机
任务分解	分析上面的工作情境，将任务分解如下： （1）查看默认虚拟主机； （2）新增虚拟主机； （3）appBase部署Java应用； （4）docBase部署Java应用
任务准备	完成Tomcat默认配置文件配置

📝 任务目标

知识目标	了解配置Tomcat虚拟主机的必要性
技能目标	学会如何配置Tomcat虚拟主机
素质目标	耐心与细心：针对输入指令报错问题进行细心的分析，配置Tomcat虚拟主机的相关指令能够耐心地完成

🔧 任务实现

步骤1： 查看默认虚拟主机。

Tomcat和Apache、Nginx一样，都支持虚拟主机配置。每个虚拟主机配置文件就是一台服务器，支持一个IP下访问多个域名，每个域名代表一个网站。所以一台服务器上可以配置多个网站。

视频

配置Tomcat
虚拟主机1

编辑Tomcat配置文件，直接在配置文件内搜索Host，找到如下内容：

```
[root@tomcat ~]# vim /usr/local/tomcat/conf/server.xml
......
  <Host name="localhost"  appBase="webapps"
      unpackWARs="true" autoDeploy="true">

  <!-- SingleSignOn valve, share authentication between web applications
      Documentation at: /docs/config/valve.html -->
  <!--
      <Valve className="org.apache.catalina.authenticator.SingleSignOn" />
```

```
-->
<!-- Access log processes all example.
     Documentation at: /docs/config/valve.html
     Note: The pattern used is equivalent to using pattern="common" -->
<Valve className="org.apache.catalina.valves.AccessLogValve"
directory="logs"
     prefix="localhost_access_log" suffix=".txt"
     pattern="%h %l %u %t "%r" %s %b" />
  </Host>
...
```

其中<Host>和</Host>之间的配置为虚拟主机配置部分。参数说明如下：

- name定义域名；
- appBase定义应用的目录；
- unpackWARs：是否自动解压war包；
- autoDeploy：如果此项设置为true，表示Tomcat服务处于运行状态，能够检测appbase下的文件，如果有新的Web应用加入进来，会自动发布这个Web应用。

Java的应用通常是一个JAR的压缩包，只需要将JAR的压缩包放到appBase目录下即可。Tomcat默认页其实就是在appBase目录下面，不过是在其子目录ROOT中。

步骤2： 新增虚拟主机。

新增虚拟主机，编辑server.xml，在</Host>下面增加以下内容：

```
[root@tomcat ~]# vim /usr/local/tomcat/conf/server.xml
......
<Host name="www.123.cn" appBase=""
          unpackWARs="true" autoDeploy="true"
          xmlValidation="false" xmlNamespaceAware="false">
    <Context path="" docBase="/data/wwwroot/123.cn/" debug="0" reloadable="true" crossContext="true"/>
  </Host>
```

docBase参数用来定义网站的文件存放路径。如果不定义，默认在appBase/ROOT下。定义docBase之后就以该目录为主了，其中appBase和docBase可以相同。在这一步操作过程中很多人会遇到过访问404的问题，其实就是因为docBase没有正确定义。

appBase为应用存放目录（实际上是一个相对路径，相对于 /usr/local/tomcat/ 路径），通常是需要把war包直接放到该目录下面，它会自动解压成一个程序目录。

搭建Tomcat之后，想要使用Tomcat访问一个网站，首先应用不能是一个传统的目录（Apache、Nginx访问网站，首先需要指定一个目录，目录里存放着PHP文件或者是HTML文件，然后去访问），Tomcat需要提供一个war包，即一个压缩包，这个压缩包中包含着运行这个网站的一些文件，包括配置、JS代码、数据库相关内容等，它们都需要打包成

war文件，且需要放置到 webapps 中。

步骤3: appBase部署Java应用。

下面通过部署Java应用来体会appBase和docBase目录的区别。

为了方便测试，下载zrlog（Java语言编写的blog站点应用，轻量；下载地址：http://dl.zrlog.com/release/zrlog-1.7.1-baaecb9-release.war）。zrlog实际就是一个war包。

配置Tomcat虚拟主机2

暂时将zrlog的war包下载到/usr/local/src目录下，命令如下：

```
[root@tomcat ~]# cd /usr/local/src/
[root@tomcat src]# wget http://dl.zrlog.com/release/zrlog-1.7.1-baaecb9-release.war
...
[root@tomcat src]# ls|grep zrlog
zrlog-1.7.1-baaecb9-release.war
[root@tomcat src]#
```

appBase支持自动解压，所以直接将war包复制到/usr/local/tomcat/webapps/目录下：

```
[root@tomcat src]# cp zrlog-1.7.1-baaecb9-release.war /usr/local/tomcat/webapps/
[root@tomcat src]# ls /usr/local/tomcat/webapps/
docs        host-manager   ROOT    zrlog-1.7.1-baaecb9-release.war
examples    manager        zrlog-1.7.1-baaecb9-release
//将war包复制到/usr/local/tomcat/webapps/目录下之后，包会自动解压（前提是Tomcat正常启动）
```

重命名war包的文件名，命令如下：

```
[root@tomcat src]# cd /usr/local/tomcat/webapps/
[root@tomcat webapps]# mv zrlog-1.7.1-baaecb9-release zrlog
[root@tomcat webapps]# ls
docs    examples   host-manager    manager    ROOT    zrlog
zrlog-1.7.1-baaecb9-release.war
[root@tomcat webapps]# ls
docs        host-manager   ROOT    zrlog-1.7.1-baaecb9-release
examples    manager        zrlog   zrlog-1.7.1-baaecb9-release.war
// 一旦重命名或删除war解压后包的文件时，war包就会再解压出一个文件夹
```

用浏览器访问http://192.168.174.155/zrlog（注意IP），如图6-2-2所示。

图 6-2-2 用浏览器访问 http://192.168.174.155/zrlog

出现安装向导，这是一个配置数据库的过程。

之前该计算机安装了LNMP，因此不再演示MySQL的安装方式，直接调用。登录数据库，在数据库中创建一个zrlog数据库和zrlog用户。命令如下：

```
[root@tomcat webapps]# /usr/local/mysql/bin/mysql
Welcome to the MySQL monitor.  Commands end with ; or \g.
Your MySQL connection id is 1
Server version: 5.6.39 MySQL Community Server (GPL)
Copyright (c) 2000, 2018, Oracle and/or its affiliates. All rights reserved.
Oracle is a registered trademark of Oracle Corporation and/or its
affiliates. Other names may be trademarks of their respective
owners.
Type 'help;' or '\h' for help. Type '\c' to clear the current input statement.
mysql> create database zrlog;
//创建zrlog数据库
Query OK, 1 row affected (0.00 sec)

mysql> grant all on zrlog.* to 'zrlog'@127.0.0.1 identified by '000000';
Query OK, 0 rows affected (0.05 sec)
//创建zrlog用户
mysql>exit
Byebye
//退出数据库
```

检查创建用户是否可以登录数据库，使用zrlog用户登录。命令如下：

```
[root@tomcat webapps]# /usr/local/mysql/bin/mysql -u zrlog -h 127.0.0.1 -p 000000
Warning: Using a password on the command line interface can be insecure.
Welcome to the MySQL monitor.  Commands end with ; or \g.
```

```
Your MySQL connection id is 2
Server version: 5.6.39 MySQL Community Server (GPL)

Copyright (c) 2000, 2018, Oracle and/or its affiliates. All rights reserved.

Oracle is a registered trademark of Oracle Corporation and/or its
affiliates. Other names may be trademarks of their respective owners.

Type 'help;' or '\h' for help. Type '\c' to clear the current input statement.

mysql> show databases;
//检查已有数据库
+--------------------+
| Database           |
+--------------------+
| information_schema |
| test               |
| zrlog              |
+--------------------+
3 rows in set (0.00 sec)
mysql> exit
Byebye
```

检查完成，zrlog用户登录成功。使用 zrlog用户信息填写刚才在浏览器中打开的网页，E-mail填写自己的邮箱，本次是实验，填写内容为自定义邮箱（tomcat@163.com），如图6-2-3所示。单击"下一步"按钮。

图 6-2-3　填写数据库信息

设置管理员账号（admin）和管理员密码（123456），网站标题和子标题按需填写，本次自定义内容（网站标题为"测试"，网站子标题为linux），如图6-2-4所示。

图 6-2-4　填写网站信息

填写完成后，单击"下一步"按钮，可以看到安装完成界面，如图6-2-5所示。

图 6-2-5　安装完成

单击"点击查看"按钮，就可以进入搭建好的zrlog页面了，如图6-2-6所示。

图 6-2-6　进入搭建好的 zrlog 页面

也可以进入管理页面，单击图6-2-6主菜单栏中的"管理"按钮，进入管理登录页面，如图6-2-7所示。

图 6-2-7　进入管理登录页面

输入安装向导中设置的用户名和密码（admin，123456），单击"登录"按钮，登录成功，如图6-2-8所示。

图 6-2-8　登录成功页面

单击"文章撰写"栏目，输入自己想写的内容，如图6-2-9所示，然后保存。

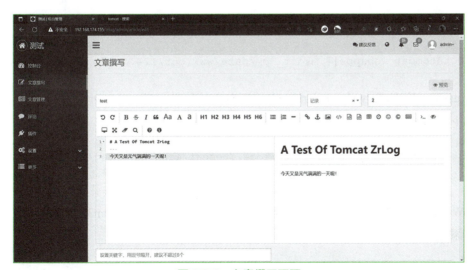

图 6-2-9　文章撰写页面

回到主页面，即可看到刚才所写的内容，如图6-2-10所示。

图 6-2-10　回到主页面

步骤4： docBase部署Java应用。

在浏览器访问zrlog时需要指定IP地址加目录，那么如何才能输入IP直接访问该目录呢？

查看虚拟主机docBase的路径，结果如下：

```
[root@tomcat ~]# vim /usr/local/tomcat/conf/server.xml
...
<Host name="www.123.cn" appBase=""
        unpackWARs="true" autoDeploy="true"
        xmlValidation="false" xmlNamespaceAware="false">
        <Context path="" docBase="/data/wwwroot/123.cn/" debug="0"
        reloadable="true" crossContext="true"/>
</Host>
//docBase 定义的目录为/data/wwwroot/123.cn
```

创建docBase目录，命令如下：

```
[root@tomcat webapps]# mkdir -p /data/wwwroot/123.cn/
```

将/usr/local/tomcat/webapps/zrlog 中的所有文件移动到/data/wwwroot/123.cn/目录下，命令如下：

```
[root@tomcat webapps]# mv /usr/local/tomcat/webapps/zrlog/* /data/wwwroot/123.cn/
```

在Windows下绑定hosts文件，hosts文件路径为C:\Windows\System32\drivers\etc，在文件下面添加如下命令：

```
192.168.174.155 www.123.cn
```

测试：

打开命令提示符（CMD），用 ping www.123.cn 命令查看 IP 是否为虚拟机 IP，如果是，即可访问，如图 6-2-11 所示。

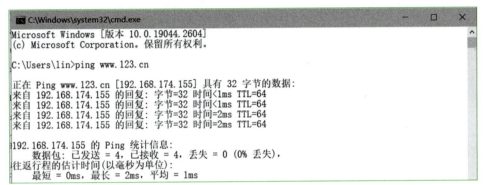

图 6-2-11　查看 IP 是否为虚拟机 IP

用域名访问zrlog页面，由于之前配置完虚拟主机后并没有重启服务，因此这里要重启服务。命令如下：

```
[root@tomcat webapps]# /usr/local/tomcat/bin/shutdown.sh
Using CATALINA_BASE:   /usr/local/tomcat
Using CATALINA_HOME:   /usr/local/tomcat
Using CATALINA_TMPDIR: /usr/local/tomcat/temp
Using JRE_HOME:        /usr/local/jdk1.8
Using CLASSPATH:       /usr/local/tomcat/bin/bootstrap.jar:/usr/local/tomcat/bin/tomcat-juli.jar
[root@tomcat webapps]# /usr/local/tomcat/bin/startup.sh
Using CATALINA_BASE:   /usr/local/tomcat
Using CATALINA_HOME:   /usr/local/tomcat
Using CATALINA_TMPDIR: /usr/local/tomcat/temp
Using JRE_HOME:        /usr/local/jdk1.8
Using CLASSPATH:       /usr/local/tomcat/bin/bootstrap.jar:/usr/local/tomcat/bin/tomcat-juli.jar
Tomcat started.
[root@tomcat webapps]#
```

重启Tomcat服务之后，打开浏览器，在地址栏中输入www.123.cn，即可看到zrlog首页，如图6-2-12所示。

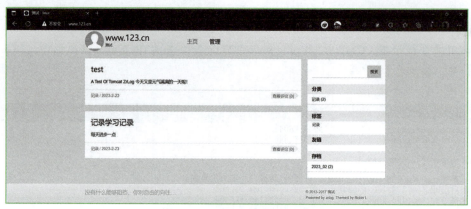

图 6-2-12　zrlog 首页

这个页面与之前用指定IP加目录访问的页面是一样的。

学习笔记

考评记录

姓名		完成日期	
序号	考核内容	标准分	评分
1	查看默认虚拟主机	25	
2	新增虚拟主机	25	
3	appBase 部署 Java 应用	25	
4	docBase 部署 Java 应用	25	
总评分		100	

任务实现心得：

任务实训

任务实训	独立完成配置 Tomcat 虚拟主机的任务
任务目标	掌握配置 Tomcat 虚拟主机的原理

任务 3　配置 Tomcat 生成日志

任务描述

视频
配置Tomcat日志

情境描述	日志在公司的日常开发中是非常重要的，程序员B就需要配置Tomcat的日志。Tomcat在应用过程中，难免会出现错误，要查看这些错误，就需要查看Tomcat的日志。接下来学习如何配置生成日志与查看日志
任务分解	分析上面的工作情境，将任务分解如下： (1) 配置生成日志； (2) 查看日志
任务准备	完成Tomcat默认配置文件配置

任务目标

知识目标	了解配置Tomcat监听80端口的必要性
技能目标	学会如何配置Tomcat监听80端口
素质目标	耐心与细心：针对输入指令的报错问题进行细心的分析，配置Tomcat监听80端口的相关指令能够耐心地完成

任务实现

步骤1： 配置生成日志。

由于日志默认不会生成，因此需要在server.xml中配置一下。接下来配置新增虚拟主机的访问日志，具体方法是在对应虚拟主机的<Host></Host>中加入下面的配置（本案例用www.123.cn进行配置），编辑配置文件server.xml。命令如下：

```
[root@tomcat ~]# vim /usr/local/tomcat/conf/server.xml
...
    <Host name="www.123.cn" appBase=""
          unpackWARs="true" autoDeploy="true"
          xmlValidation="false" xmlNamespaceAware="false">
    <Context path="" docBase="/data/wwwroot/123.cn/" debug="0"
reloadable="true" crossContext="true"/>

    <Valve className="org.apache.catalina.valves.AccessLogValve."
           directory="logs"
           prefix="123.cn_access" suffix=".log"
           pattern="%h %l %u %t "%r"%s %b"/>
    </Host>
...
[root@tomcat ~]#
```

项目 6　Tomcat 搭建配置

> **说明：**
> - valve 为日志文件配置；
> - prefix 定义访问日志的前缀；
> - suffix 定义日志的后缀；
> - pattern 定义日志格式。

> **注意：**
> 新增加的虚拟主机默认并不会生成类似默认虚拟主机的那个 localhost.日期.log 日志，错误日志会统一记录到 catalina.out 中。关于 Tomcat 日志，最需要关注 catalina.out，当出现问题时，应该第一想到去查看它。

配置完成后，重启Tomcat服务。命令如下：

```
[root@tomcat ~]# /usr/local/tomcat/bin/shutdown.sh
Using CATALINA_BASE:   /usr/local/tomcat
Using CATALINA_HOME:   /usr/local/tomcat
Using CATALINA_TMPDIR: /usr/local/tomcat/temp
Using JRE_HOME:        /usr/local/jdk1.8
Using CLASSPATH:       /usr/local/tomcat/bin/bootstrap.jar:/usr/local/tomcat/bin/tomcat-juli.jar
[root@tomcat ~]# /usr/local/tomcat/bin/startup.sh
Using CATALINA_BASE:   /usr/local/tomcat
Using CATALINA_HOME:   /usr/local/tomcat
Using CATALINA_TMPDIR: /usr/local/tomcat/temp
Using JRE_HOME:        /usr/local/jdk1.8
Using CLASSPATH:       /usr/local/tomcat/bin/bootstrap.jar:/usr/local/tomcat/bin/tomcat-juli.jar
Tomcat started.
[root@tomcat ~]#
```

重启Tomcat服务完成后，访问网站，查看/usr/local/tomcat/logs目录下是否有日志生成，并且查看生成的日志信息。命令如下：

```
[root@tomcat ~]# ls /usr/local/tomcat/logs/
123_access_log.2020-03-29.txt    localhost.2020-03-29.log
catalina.2020-03-29.log          localhost_access_log.2020-03-29.txt
catalina.out                     manager.2020-03-29.log
host-manager.2020-03-29.log
[root@tomcat ~]# cat /usr/local/tomcat/logs/123_access_log.2020-03-29.txt
192.168.174.1 -- [29/Mar/2020:22:06:16 -0700] "GET / HTTP/1.1" 200 8645
192.168.174.1 -- [29/Mar/2020:22:06:16 -0700] "GET //favicon.ico
```

6-27

```
HTTP/1.1" 200 9662
[root@tomcat ~]#
```

至此，Tomcat搭建配置完成。

步骤2: 查看日志。

Tomcat日志存放在/usr/local/tomcat/logs/目录下。命令如下：

```
[root@tomcat ~]# ls /usr/local/tomcat/logs/
catalina.2020-03-29.log      localhost.2020-03-29.log
catalina.out                 localhost_access_log.2020-03-29.txt
host-manager.2020-03-29.log  manager.2020-03-29.log
[root@tomcat ~]#
```

> **说明：**
> • catalina 开头的日志为 Tomcat 的综合日志，它记录 Tomcat 服务相关信息，也会记录错误日志。
> • catalina.2020-xx-xx.log 和 catalina.out 内容相同，前者会每天生成一个新的日志。
> • host-manager 和 manager 为管理相关的日志，其中 host-manager 为虚拟主机的管理日志。
> • localhost 和 localhost-access 为虚拟主机相关日志，其中带 access 字样的日志为访问日志，不带 access 字样的为默认虚拟主机的错误日志。

学习笔记

考评记录

姓名		完成日期	
序号	考核内容	标准分	评分
1	配置生成日志	60	
2	查看日志	40	
	总评分	100	

任务实现心得：

任务实训

任务实训	自行独立配置 Tomcat 生成日志与查看日志
任务目标	掌握配置 Tomcat 生成日志与查看日志的方法

参 考 文 献

[1] 苏军. Docker容器安全管控技术研究[J]. 网络安全技术与应用，2020(11)：21-22.

[2] 吕彬，徐国坤. Docker容器安全性分析与增强方案研究[J]. 保密科学技术，2021(1)：15-22.

[3] 陈金窗，沈灿，刘政委. Ansible自动化运维：技术与最佳实践[M]. 北京：机械工业出版社，2016.

[4] 李松涛，魏巍，甘捷. Ansible权威指南[M]. 北京：机械工业出版社，2016.

[5] 裘紫阳. 面向Elasticsearch的监控管理平台的设计与实现[D]. 武汉：华中科技大学，2019.

[6] 徐伟杰，王挺，薛婉婷. 基于Elasticsearch的搜索引擎设计与实现[J]. 智库时代，2019(23)：228，240.

[7] 冯冬艳. MySQL集群主从复制的原理、实现与常见故障排除[J]. 山西电子技术，2021(6)：47-48，57.

[8] 李艳杰. MySQL数据库下存储过程的设计与应用[J]. 信息技术与信息化，2021(1)：96-97.